Audio-Visual Equipment

Audio-Visual Equipment

A technician's and user's handbook

Ian Robertson

Newnes
An imprint of Butterworth-Heinemann Ltd
Halley Court, Jordan Hill, Oxford OX2 8EJ

 PART OF REED INTERNATIONAL BOOKS

OXFORD   LONDON   GUILDFORD   BOSTON
MUNICH   NEW DELHI   SINGAPORE   SYDNEY
TOKYO   TORONTO   WELLINGTON

First published 1991

British Library Cataloguing in Publication Data
Robertson, Ian
    Audio-visual equipment.
    1. Audiovisual equipment
    I. Title
    621.3897

ISBN 0 7506 0021 7

Typeset in Great Britain by Saxon Printing Ltd, Derby.
Printed and bound in Great Britain.

Contents

Preface

The increase in the use of audio-visual (AV) methods over the past fifteen years in schools and colleges has made considerable demands both on technician training and on supply of equipment. It has also had a considerable impact on suppliers of spare parts, because much AV equipment in schools and colleges is old and requires considerable maintenance, not least because of the amount of handling – often not particularly sympathetic – it experiences. Teacher-training courses also must reflect the AV use, because the correct use of AV requires some knowledge of the technology and considerable experience. AV use is also widespread, almost universal, in conference centres and in places such as large hotels in which conferences and lectures are convened. In addition, AV use is now seen as a vital aid to management communications in many firms, large and small.

This very considerable growth in the use of AV has not been matched by a similar growth in the training of AV technicians, though colleges which run the City & Guilds laboratory technician courses (No.735) usually specify a period of AV training, while the 734 course is devoted entirely to audio-visual aids. One of the problems facing such courses is the breadth of work which ranges from photography to electronics servicing. Another problem is that users of AV equipment are often unfamiliar with what can be achieved, and what they can reasonably expect the AV technician to be able to do. This applies most forcefully to the business sector, in which many users of AV equipment do so only on odd occasions.

This book is an AV guide for both user and technician. For the user, it shows the range of AV equipment, how it works and what it can do. For the technician it shows the finer points of choice of equipment, how it should be maintained and the very important topic of how to prepare material. It also stresses how the AV technician can help the user to understand the equipment and avoid problems which so often mar a poorly-planned AV display. Outside educational establishments, users of AV equipment may also be required to ensure equipment is in good order, and for such readers this book will be of particular importance. It should also be of considerable help to trainee teachers and lecturers in understanding the capabilities and, equally important, the limitations of AV techniques.

No attempt has been made to describe servicing procedures other than routine maintenance in full, because to do so would require a set of volumes of encyclopaedic proportions. One minor exception has been made for 16 mm projectors, for which detailed technical manuals are not easy to come by. In general though, the large range of types of such equipment and age of equipment makes it impossible to cover more than the essentials. In any case, more modern equipment, such as video equipment, requires specialized servicing for which few audio-visual aids (AVA) departments are equipped. So by concentrating only on items which a non-specialist technician can be expected to attend to in order to keep equipment in good order, over 90% of the work required of AVA departments is covered by this book.

I am most grateful for the help received in the preparation of this book. Above all, I owe a debt of gratitude to Alan Hambly, formerly chief AV technician at Braintree College of Further Education, who introduced me to AV methods in the first place.

I am also most grateful to the firms listed here, which unstintingly supplied information on AV products and whose help was fundamental to this book: APT Radar Systems (Cybervox Division); De Vere; Elf Holdings; Eric Fishwick; Folex; Hewlett Packard; Kodak; Konica Business Machines; Linguaphone Group; Matmos; Rotaprint; RS Components; Sony; Star Micronics (UK).

Ian Robertson

1

The overhead projector

The overhead projector (OHP), though not the oldest of visual aids, has become the most central and indispensable and is now the means of displaying live computer screens (as distinct from showing transparencies taken from screens) to large audiences (see later). Virtually any type of visual material can be converted into OHP transparencies, though for some, notably 35 mm slides, conversion is time-consuming and tortuous. Wide availability of OHP in schools, colleges and conference centres ensures that any speaker with a stock of OHP transparencies should always be able to find a means of displaying illustrations.

Many modern types of OHP fold for easier carrying, and the greater use of plastics to replace steel makes modern OHPs considerably lighter than older versions. Other features of modern OHPs include almost-silent fans, instant bulb-change facilities, interchangeable lens heads, roll cassettes to protect transparency rolls against dust, and fold-down facilities which allow 35 mm slides also to be projected. This latter facility is extremely useful where both slide and transparency material is to be displayed.

Principles

Principles and light-path of a typical OHP are illustrated in Figure 1.1. The transparency, typically of A4 size, has to be evenly illuminated so the light-source must be either a flat-panel source or (more usually as shown) a conventional projector bulb and collimator lens. A collimator lens (also sometimes called a condenser lens) is one arranged with a light-source at its focus, so that emerging light consists, as nearly as possible, of parallel beams. Size of the lens required for OHP is as large as the transparency

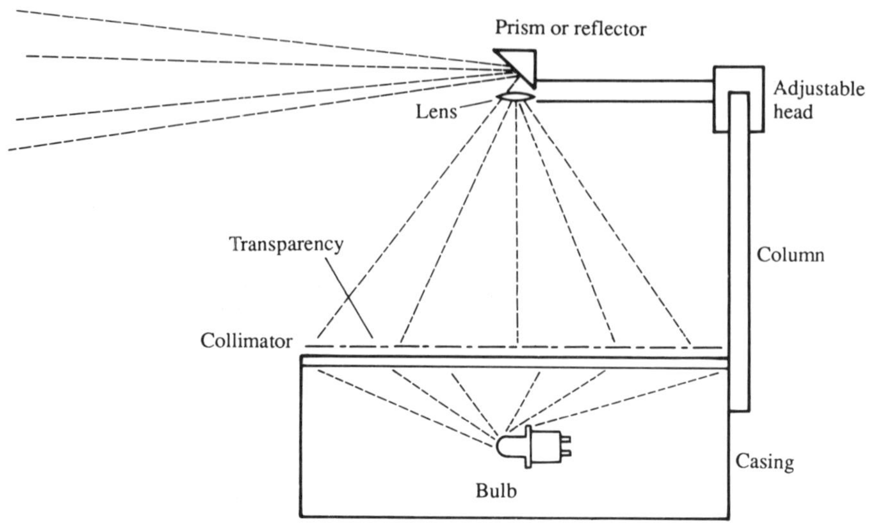

Figure 1.1   *Main parts and light-path of a typical modern overhead projector (OHP)*

itself which rules out, on grounds of size and price, convex glass lenses used in filmstrip and slide projectors. The collimator for the OHP is, in fact, a Fresnel lens or zone plate using the principle discovered around 1818, in which a set of opaque circles (Figure 1.2) drawn on a transparent surface can act as a lens. Circles are arranged so their spacing decreases as their radii increase. Equivalent focal length of a Fresnel lens depends on radii spacings, according to the expression:

$$r = \sqrt{nf\lambda}$$

where **r** is the radius of the circle, **n** is an even number, **f** is the focal length required and λ is the wavelength of light.

If a value for λ for yellow light is used and a focal length of 100 mm is assumed then with dimensions in millimetres the expression simplifies to $r = 0.2345 \times \sqrt{n}$, using values for n of 2, 4, 6, 8 and so on to find radii of successive circles.

It is possible, therefore, to make a transparency which is a copy of a Fresnel lens pattern. Such a transparency allows an OHP to be used if by some mischance its main Fresnel lens has been shattered. Fresnel transparencies also have many other applications, notably the lens stuck to the rear window of a bus allowing its driver to see a wide-angle view of anything behind the bus not in the normal line of rear vision through the mirror. This

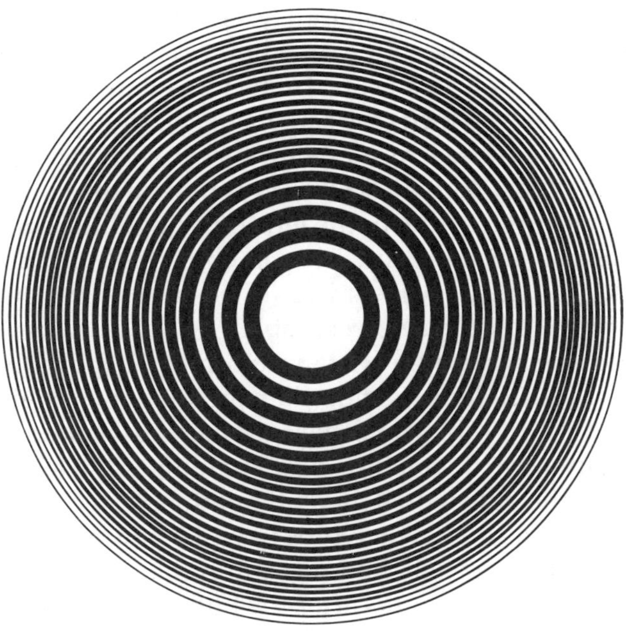

Figure 1.2   *Fresnel lens or zone plate – consists of a set of concentric circles, each with a radius whose value is found from a formula*

application requires a Fresnel lens with a much longer focal length than that used for OHP.

The large transparency area illuminated requires a projector lamp of considerable power, particularly as the OHP is often used in daylight conditions or full room lighting, rather than in the total blackout needed for slide or movie projection. The standard form of bulb is the 24 V 250 W quartz-iodine type, though OHPs in large lecture theatres are likely to use 400 W metal-vapour discharge lamps, which feature a light output considerably more than that of the same electrical power of incandescent lamp. On smaller OHPs smaller filament size of quartz-halogen bulbs greatly assists collimation and efficient conversion of electrical power into light intensity – more so than was ever possible using older types of projection lamps. Bulb power makes it essential to use a fan to keep the area round the lens reasonably cool, and fan failure should automatically shut off power to the bulb.

The projection part of the OHP consists of a projection lens, usually of large diameter and focal length of 260 to 300 mm, together with a prism or mirror giving a 90° change in light direction. It is this combination of lens and mirror which ensures images on screen and transparency are the same way up. The projection head can be tilted to allow for different heights of screen, and the whole head can also be moved up or down relative to the transparency to focus the screen image, though a few models use a movement of the lens alone for focusing. Lenses are usually fairly simple types, so that black-and-white images on the screen have coloured fringes, due to the lens focusing different colours to different places (dispersion). This is correctable by using more elaborate lenses or a concave mirror in place of a lens, but for most purposes fringe colours are acceptable.

The switching system allows fans to run while the bulb is unlit so a (plastic) Fresnel lens is not damaged by the high residual temperature of a bulb which has just been switched off. Many users find fan noise distracting, and are tempted to switch the fan off whenever the bulb is switched off but this should be avoided. A better system is the one-switch type in which the fan is switched on whenever the bulb is switched on, and remains running (using a time delay) for a few minutes after the bulb has been switched off. In any case, modern OHPs use fans with noise levels much less than older varieties. A noisy fan on such a machine indicates the need for an overhaul.

User guide

The OHP user has to ensure (1) material to be projected is suitable for OHP use, (2) the OHP is used correctly, and (3) set-up of screen and room lighting is suitable. In particular, a screen should not be placed where a spotlight illuminates it or, if such a position is unavoidable, the spotlight should be switched off when the OHP is being used. When projection is on a whiteboard, or a screen placed over a blackboard, it is quite usual to find one or more spotlights positioned to illuminate this area brightly.

The main problem with screen positioning is keystone distortion. The ideal arrangement of OHP and screen is shown in Figure 1.3, where screen surface is at right angles to the centre of the light beams. This is seldom possible, because such an arrangement either places the OHP awkwardly for the user, or has the screen sited too low for the viewers. The usual response is to place the screen higher and tilt the projection head (Figure 1.4) but this causes keystone distortion of the image due to the top of the screen being at a different distance from the projector head relative to the bottom. This makes the image of a rectangle distort as illustrated in Figure 1.5 with its lower edge narrower than its upper edge, and all other images are similarly distorted. For some types of visual material

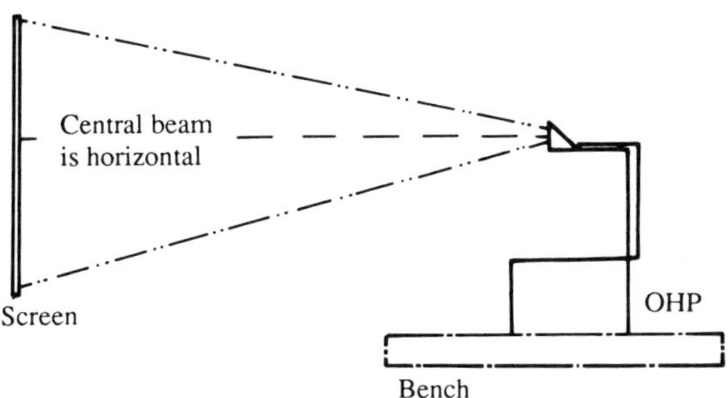

Figure 1.3   *Ideal relationship of projector and screen, with the screen at right angles to the central ray from the projector*

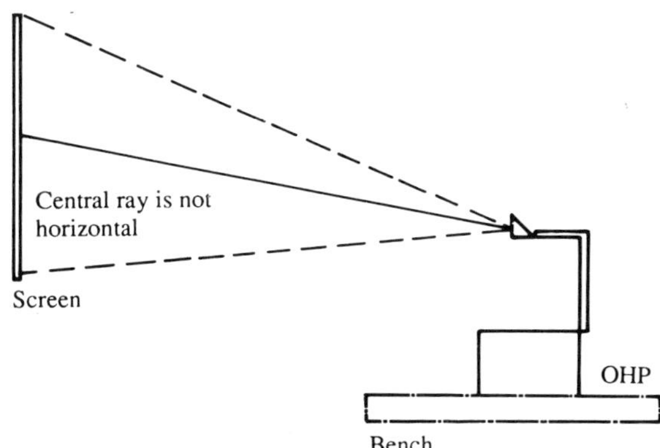

Figure 1.4   *With the screen placed high and the projector tilted, images suffer from keystone distortion, caused by unequal distance of rays to different parts of the screen*

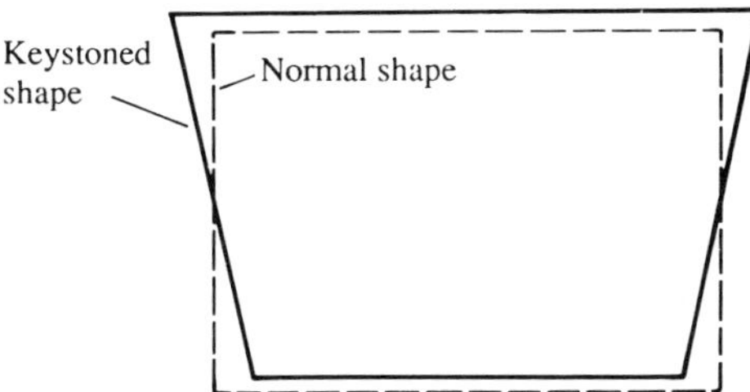

Figure 1.5   *Typical keystone effect, showing shape of a rectangle affected by keystone distortion*

this is not particularly noticeable (for abstract art patterns it may even be a bonus!) but where a projected image contains geometrical patterns (a rectangular frame round a picture, for example) or text, the effect can be very objectionable. The greater the tilt of projection head, the worse the distortion. Keystone distortion is most difficult to eradicate when wide-angle lenses are used to display large-area images in small rooms.

The cure for keystone distortion lies in restoring the angle between beam centre and screen surface to 90°. Increasing the distance between OHP and screen makes some improvement, but by far the better method is to tilt the screen (Figure 1.6) so the angle between beam and screen is correct. Screens specifically designed for use with an OHP usually tilt but whiteboards and screens intended for use with slide or movie projectors do not. Worse still, when a screen consists of a cloth or plastic surface, tilting the screen often causes the material to sag into a convex shape, causing even worse distortion than by the keystone effect. Such screens should be tilted only if the material can be kept taut and the surface reasonably flat. Another advantage, incidentally, of tilting a screen is the consequent reduction of the ambient illumination on the screen (unless you are working on a stage with footlights) and the relative increase of brightness of the displayed image.

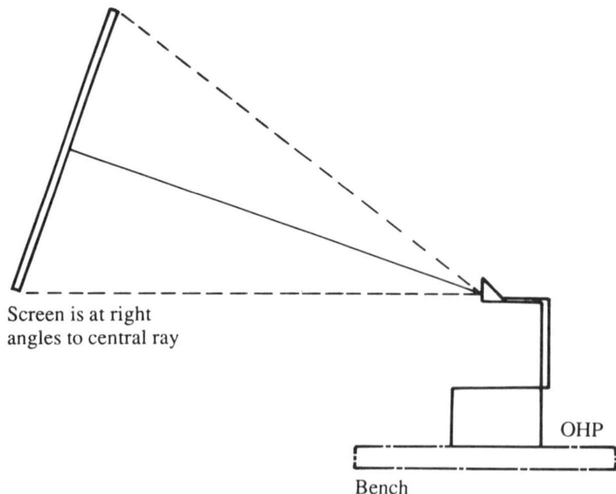

Figure 1.6   *Correcting keystone distortion by tilting the screen, making the central ray once again at right angles to the screen surface*

Incidentally, keystone distortion will also be present if the screen is displaced laterally from the OHP so the beam is aimed at an angle.

Horizontal keystoning results in a picture whose sides are of unequal proportions, and is just as objectionable as the more familiar vertical keystoning. The remedy is to turn the screen or move the projector so a 90° angle is restored.

When the OHP is in use, try to avoid looking into the light beam. This can be difficult if you need to lean over the illuminated transparency to emphasize some aspect of the image, but it is usually possible to use a short pointer without getting your head over the illuminated area. It is equally important not to look into the projection lens, because the beam close to the lens is very bright, and it will be some time before your vision returns to normal (by which time the average school class will have run amok). The surface of the Fresnel lens is also bright, and models of OHP which feature an anti-glare coating for the Fresnel lens are noticeably easier on the eyes. One useful point to remember is that glare is much less prominent if you are working in good lighting – avoid having the OHP standing in a pool of darkness.

OHP movement should be done carefully and while the bulb is cold. Early types of projection bulbs were quite fragile and required careful handling of the OHP. Though bulbs are now much more robust they still cannot tolerate careless handling, particularly setting the machine down sharply. Even if a bulb survives in a roughly handled OHP, its working life is inevitably reduced and it is a matter of practical observation that a bulb always fails just as it is switched on for an important presentation, never when the technician or user tests it beforehand. Bulbs also display a tendency to fail when the technician has gone for tea, leaving the spares cupboard locked. Many modern machines feature a bulb-change mechanism, allowing a second bulb to be moved into position and switched on without the need to open the machine or touch the bulb. This scheme works only if users and technicians co-operate, so that users report bulb failure and technicians ensure both bulbs are working before the machine is used. If this is not done, odds are a bulb will blow on a machine which already contains one defective bulb.

Preparation

Preparation of material for the OHP is a task usually shared between user and technician, with the user responsible for origins and choice of materials and the technician responsible for preparing it for transparency use.

The simplest scheme of OHP use is to draw directly on the transparent roll of acetate film usually provided with the OHP or to draw on separate sheets which can be kept in folders. Use of rolled film is often restricted to impromptu sketches which are not needed again and which will be washed off the roll when the whole of the roll has been used. When separate sheets are used, drawings are usually of a higher standard, intended to be used more than once. Such sheets may be placed under the rolled film allowing notes, arrows and other items to be drawn without making any change to the single sheet underneath. An effective development of this principle is the use of overlays in which an image is built up from several sheets of transparencies, each of which adds another layer of detail. Bought-in transparencies are usually mounted in holders, and OHPs generally feature a set of (retractable) locating pins to ensure transparencies are correctly positioned.

Overlays are particularly effective when a class is expected to make a sketch. Few users of the OHP allow for the restricted artistic abilities of those watching and if a pattern can be shown as a few lines at first, with more detail added in each overlay, the whole image becomes much simpler to draw. This is one of the advantages to the use of a blackboard by a good practitioner – each stage in a drawing is shown step-by-step, making the whole less intimidating. OHP use has tended to lead towards displays of elaborate line drawings and photographs which are very difficult to sketch with any accuracy. Despite what some educationalists say, visual images are not necessarily well-remembered, particularly in these days of instantly-forgettable television, and making a sketch of your own is a much more positive aid to recalling the image. It also keeps a class occupied; a useful point to bear in mind.

Another advantage of using overlays is their use for a form of animation. One transparency contains the non-changing lines of a drawing, while the overlays contain successive stages of movement of a mechanism or other action. By retaining the background transparency and changing foreground overlays action can be seen in a way not always apparent from a still. Many still pictures, for example, do not show particularly well which parts will move and in which direction movement will be. In addition, many viewers who can visualize which parts move and where hinges or fulcrums are placed nevertheless have great difficulty in imagining the movement. Oddly enough, viewers who are studying the imaginative arts usually have these difficulties, not the supposedly unimaginative engineers. An advance on the use of successive-position drawings is the transparent model on which parts can be moved to show working principles.

The greatest difficulty about drawings is using suitable media. Many felt-tipped pens sold for OHP use are coarse-tipped, consequently drawings can consist of coarse outlines only. Certainly a drawing should not contain too much detail, but when a pen permits only one or two words to be written across the display (as some do) then the OHP becomes a hindrance rather than an aid. At the other extreme, some pens whose tips are fine enough for good drawing work make lines which are too transparent and difficult to see when projected. It is not difficult to make a good choice of pens but when such items are bought in bulk by, say, a local authority they are more likely to be bought because of their price, because of blandishments by a representative of the manufacturer, or as part of a package deal. Unsuitable pens are one of the most common causes of OHP machines being left stacked, unused, in a cupboard.

For some types of work, colour is useful to fill in areas or to differentiate between different sections. This requires two sets of colour pens, one more opaque to be used for coloured boundary lines, and the other set more transparent, to be used for filling in. The two should be kept apart and well-labelled, because considerable frustration can result from trying to draw with a pen whose traces are almost invisible or from filling in an area and then finding that it looks almost black when projected.

Many users of OHPs never draw, but rely on reproduction of other graphics work. This involves making a selection of drawings and photographs and relying on the visual aids technician to prepare suitable transparencies. This makes the life of the user considerably easier, but raises the question of violation of copyright. There is an agreement negotiated through ALCS (Authors Collecting and Licensing Service) for the photocopying of text, by which a small copying fee can be distributed to authors whose work (usually sociological or technical books), has been copied and distributed. No such agreement seems to cover the work of artists whose painstakingly drawn illustrations are copied and exhibited on the OHP, but it is unlikely completely free use of such illustrations will continue following a recent case in which a recognizable painting appeared in a published photograph.

Another route to OHP master preparation is by way of the computer. A dot-matrix printer will print on to suitable (slightly matte surfaced) transparent film, and suitable printing films are also available for laser printers. Initially, this method was used only for making text masters, and suffered from the usually small print of the dot matrix printers when used with computers and software of the early 1980s. More recently, however, desk-top publishing software has become available which allows better control of text size, making text much more readable when projected. Low prices of colour inkjet printers make these an attractive option for producing coloured transparencies from computer displays too, but few types of transparency material take the coloured inks without smearing. Thermal colour copiers are expensive but can turn out excellent work. A considerable advantage of making copies of a computer screen as distinct from displaying the screen itself (see later) is that time lags within a presentation are avoided. For example it might take a computer several minutes to draw a bar-graph display from a spreadsheet, but if this is transferred to a transparency it can be shown whenever needed.

A good guide here is to use text whose size is at least enough to make 40 characters per line, although 30 characters per line is better. Many schools are, unfortunately, tied by local authority agreements to computers which are not IBM-compatible and so do not have access to the vast amount of low-cost software (and low-cost computers and peripherals) which the use of such machines can bring. Technical colleges and independent schools are less restrained and commercial enterprises will, in any case, be using compatible machines.

The use of IBM-compatible machines also allows drawings to be made using CAD (computer aided design) software. This is of enormous assistance in preparing technical drawings, and the simpler packages can be quickly mastered by users who have done little more than prepare pencil sketches previously. Drawings prepared in this way print out to as high a standard as the printer itself will permit. This is in contrast to drawing software, in which the computer screen determines the appearance of lines so that unless a very high-resolution display is used, diagonal lines always look very jagged. Virtually any modern dot-matrix printer can produce, using a CAD package, diagonal lines looking almost as good as any produced with a ruler and a felt-tipped pen. Using a laser printer along with CAD results in work whose quality cannot be faulted. Non-CAD drawing and painting software is more limited because of its poor appearance when printed and projected, unless a high-resolution screen display is used. In any case, most painting packages are geared to the use of colour screens, and lose their impact considerably when printed in monochrome.

Though colour printers are available, there is no guarantee a colour printer will produce acceptable results with a colour painting software package. One impressive package catering for many colour printers is *Deluxe Paint II*, for IBM-compatible computers.

The most direct form of use of the computer for OHP displays is the use of the LCD panel display. This is a comparatively thin computer monitor screen which uses the LCD form of shadow display rather than the light-emitting cathode-ray tube of the conventional monitor. The OHP illumination system then acts as the light source, with the LCD screen being used as the transparency, connected to the monitor output of the computer and displaying whatever the computer would normally be displaying. This allow the user to project a computer image on whatever size of screen is available, without the complications, size, weight and cost of a large-screen monitor.

LCD panels, such as are available from *Folex* and other manufacturers, can be obtained in monochrome or colour versions, and are, as would be expected, intended to be connected to IBM-compatible machines (a range which takes in *Amstrad* and *Dell* among several dozens of suppliers). This excludes many of the incompatible machines used in schools, and as the screen is part of the hardware of the computer system, users cannot get round the incompatibility problem by using software emulators which allow schools computers to run IBM software.

Quoting specifications of Folex panels, the monochrome type is intended to be used with CGA screen displays. Machines which use the later EGA or VGA screen displays are almost always CGA-compatible, so that they can make use of this panel if they are switched (usually by software commands) to CGA but the monochrome-only type of screen, such as the Hercules, cannot. Output from the computer should be of the RGB type even for the monochrome display.

Resolution for the monochrome LCD panel matches that of the CGA monochrome screen, 640 × 200 pixels. The word pixel is an abbreviation of the term *picture element*, and means the number of picture dots whose brightness can be controlled independently. The word *dot* is not particularly appropriate here, because each element is a rectangle whose vertical height is considerably more than its width. For example, if the screen dimensions are 230 mm wide by 170 mm high, each *dot* on a 640 × 200 display will be 0.36 mm wide and 0.85 mm deep, a ratio of 2.36 of depth to width. Though this resolution is rather better than a domestic TV receiver, the 200 dots vertical resolution means diagonal lines look very jagged, particularly on a large display. Even on paper, 200-pixel vertical resolution implies about 25 dots per inch, well below the 100 dots per inch that was once used for newspaper reproduction of photographs and the 300 dots per inch of laser printers (or the 1200 dots per inch used for modern typesetting from computer). Absence of colour is not necessarily a great restriction, because most of the software packages used for presentations of data allow for shading of areas to distinguish one area of a drawing from another.

The colour version of the LCD panel can make use of CGA or EGA colour systems and also the excellent Hercules monochrome system which is fitted to a large number of lower-priced PC machines (in the £300 or lower price category). The Hercules display allows much higher resolution than CGA and is ideally suited to CAD work and text. The colour LCD screen, used with a colour output from the computer, permits resolution up to 640 × 400 pixels, with eight colours (counting black and transparent as colours). This allows excellent reproduction of coloured material, particularly graphics for business applications such as bar-charts and pie-charts. Cost difference between LCD panels and any large-size colour displays based on TV screen is enormously in favour of the LCD panels, and this type of application is expected to increase.

The rate of development of LCD panels is such that most small-screen TV receivers will be using this type of display within the lifetime of this book, and this will be reflected in the provision of TV display panels for OHP use, allowing large-screen TV to be used at practicable costs without the bulk and weight of the older style of equipment using cathode ray tubes.

Technician guide

Maintenance of the OHP is fairly simple compared with other AV aids, and centres on bulb replacement and cleaning. For single-bulb machines in regular use it ought to be possible to predict the failure times of bulbs, and to draw up a replacement schedule, with a note attached to each machine stating the expected replacement date for the bulb. Though it always seems wasteful to replace a bulb which is still working, this is very much cheaper than waiting for a bulb to fail, because the waste of time involved when a bulb fails in the middle of a presentation is always greater than the price of a bulb. It is not always possible to expect a local authority to take this view, however, because time does not appear on a balance-sheet.Machines which feature instant bulb replacement can make use of each bulb until it burns out, so that replacement of only the failed bulbs is required.

Replacement of bulbs should, if OHPs are not subject to rough treatment, be at fairly infrequent intervals. Modern bulbs have a long life, and even when a projector is in constant use it may not be necessary to change bulbs more than once a year. Old bulbs from single-bulb machines can be kept for testing purposes or for use in low-voltage lighting; they need not be wasted. Cleaning is something which needs to be done much more frequently. The fan is often the most grubby part of any OHP, usually because it tends to be neglected. It is particularly important to keep the fan and all the air vents around it clean and well-dusted, because fan failure quickly puts a halt to the use of the OHP even if the fan is not electrically interlocked to the bulb. The fan may also require lubrication at intervals, but the maker's recommendation should be closely followed here because lubrication in excess can turn dust into grinding paste, and plastic bearings either need no lubrication or require specialized lubricants. Always avoid use of mineral oils on plastics, because the usual effect is to make plastic materials swell. A ticking fan is annoying, and the sound is usually due to a revolving part scraping against a housing, either at the blades of the fan or at some point in the motor. The problem should not be neglected because it indicates something may be working loose.

Most obvious cleaning requirements are of the Fresnel lens and the projection lens and prism/mirror assembly. These should require no more than dusting but, at times, enthusiastic drawing work can leave felt-tipped pen marks on the Fresnel surface. These can be removed but considerable care is needed to avoid marking the (usually plastic) lens surface in the process. Manufacturers' manuals should recommend a cleaning method, but the problem is usually that the felt-tipped pens used are of types which do not use water-based colours, and the recommended treatment for the lens is soap and water. Though a Fresnel lens is fairly tolerant of marks, everything possible should be done to avoid making marks when cleaning the surface – enough marking occurs in use without adding more.

The projection system should need no more than the occasional polish with optical wiping cloth, as the usual marking is caused by fingerprints. Marks on the projection lens system are not visible (because they are out of focus) when projecting, but they cut down the amount of light available for the image. While the projection lens system is being cleaned the rack-and-pinion gearing for focus adjustment can be checked and, if permitted, lubricated.

Much maintenance work on OHPs is spent on the film rollers when the users are devoted to making drawings on the roll. When the felt-tipped pens use water-soluble ink, the transparent rolls should be cleaned with water and detergent (using only a minimum of detergent). Some makes of sponge-foam strips intended for washing domestic windows are of the correct width for rapid cleaning of transparency roll film. If several rolls have to be cleaned, it is not difficult to set up a cleaning system, consisting of rollers and foam sponge in water. The cleaning operation then amounts to no more than winding the roll from one roller to another, but it is usually necessary to dry the film, either by spreading it out, or by passing it over a dry foam sponge, as shown in Figure 1.7.

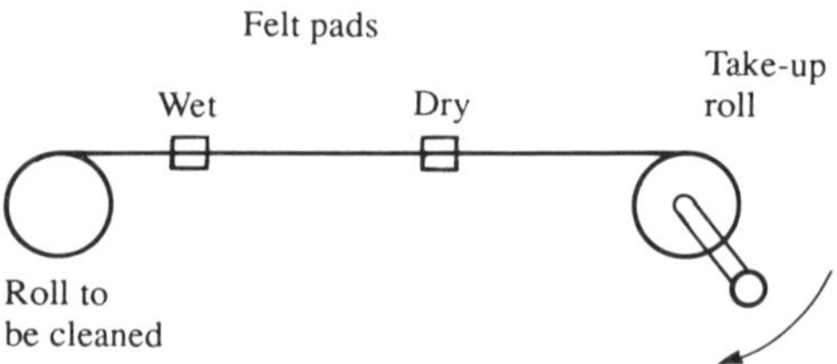

Figure 1.7   *A simple cleaning system for transport OHP rolls*

Cleaning transparency material is much more difficult when the felt-tipped pens with spirit-soluble dyes are used. Methylated spirit is usually needed to clean such rolls, and this presents a fire and vapour hazard – the spirit should be used just to dampen cloths or foam sponges and the bottle kept stoppered at all other times. Spirit-soluble felt-tipped pen should preferably be used only for permanent drawing, and water-soluble types used for work on transparent film rolls. Even when a transparent film has been cleaned after use with the spirit-based type of pens, the surface is usually stained and much less useful.

Preparation of resources

The burden of preparing transparencies for the OHP falls mainly on the visual aids technician, and requires a considerable mixture of skills. Though it is definitely not part of the normal work of the technician to make drawings for graphically-dyslexic staff, a little friendly co-operation can sometimes be very productive. Technicians similarly ought not be called on to find suitable drawings for OHP use, but at times the technician will have a better idea of where to find such material than the prospective user. This is particularly true if the technician maintains a library of visual aids material.

Conventional preparation work involves making transparencies from drawings and other material supplied by users. This requires use of photocopiers of the *Xerographic* office type (see Chapter 7), as well as thermal copiers. Older copiers work in monochrome only, but later versions produce excellent colour transparencies from a variety of sources. Office copiers often provide for enlargement and reduction facilities, but thermal copiers copy to full-size only.

Line drawings and text pages are by far the easiest type of material to prepare. Very often a simple copy to transparent film is all that is needed, with no scaling up or down required. As most book pages are appreciably smaller than the viewing surface of the OHP, however, scaling up is a distinct advantage in making the material much easier to read when projected. Text in particular benefits from being scaled up, so page width fills the width of the projection area.

Care is needed if the work involves the use of overlays, because of the need to ensure correct registration when they are placed on top of each other. It is quite common for the technician to be presented with a set of drawings photocopied from a book (sometimes from several books), all to different scales, and be instructed to make a set of overlays. This requires making a base layer to an appropriate scale, and then measuring the dimensions of the proposed overlays carefully to find what scaling needs to be used. In some cases the scaling required is beyond the capacity of the copier, and must be done in two stages, making an enlarged paper copy and subsequently a more-enlarged transparency.

The use of scaling copiers has replaced older cumbersome photographic techniques for preparing enlarged material but, as colour copiers are still fairly rare, preparation of colour material requires more work than is needed for line drawings. Most of the thermal copiers available on the market make excellent colour transparencies at a 1:1 scale but, if re-scaling is needed, photographic techniques are required (see Chapter 7). Another possibility, if high resolution is not needed, is to use a scanner connected to the computer so the image is stored in digital form, manipulated as required (altering scale, rotating, inverting, adding labels and so on) then printed back on to transparency.

Storing and cataloguing

OHP use soon results in the accumulation of a considerable amount of transparencies. Users in school and colleges generally keep their own AV materials but, in some cases and particularly when material is widely shared, the AV technician becomes responsible for storing and cataloguing this material. Though it is not a problem to keep a few transparencies in good condition, maintenance of several hundred transparencies is quite another matter because, left in a pile, transparencies stick to each other and eventually become so well attached that separation leaves traces of one image on another.

Ideally, transparencies should be held in albums, separated by tissue paper, but this is an ideal seldom achieved. One useful option is the use of two parallel rods and to punch transparencies to match rod spacing so each transparency can be hung from the rods (Figure 1.8). This makes it easier to look at each sheet, but it is not always feasible in cramped conditions to make the space needed to house the material in this way.

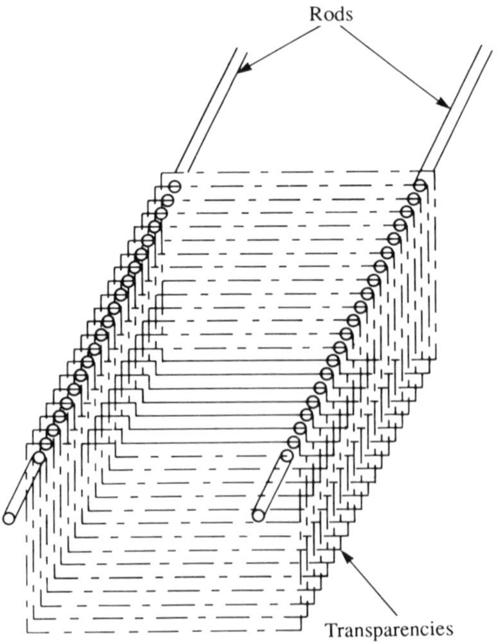

Figure 1.8   *A method of filing transparencies using two parallel rods*

Another solution is to use cheap ring-binders, with ordinary paper sheets between films. This is useful if the transparencies can be grouped into classes with the whole of one class collected into one binder and the binder labelled. This has the further advantage that the user can take the whole binder and make use of the transparencies as and when needed, rather than handling the transparencies themselves on the way from the AV resources room to the point of use.

Specialized storage books and other equipment are available, but as AV departments are often run on a proverbial shoestring many users in educational work can never hope to lay their hands on purpose-made storage units. Business users of AV equipment may benefit, however. This makes the use of the LCD panel computer display of even greater interest because the drawing is a file on the computer disk system, and large numbers of images are held in a very small space.

Cataloguing of AV material is particularly difficult because of the interlocking nature of the material. Just to give one example, a lesson on the acoustics of a stopped pipe could involve transparencies with overlays showing the pattern of sound waves in the pipe, several slides of instruments such as oboe and clarinet, a piece of movie film and a sound track for the benefit of youngsters who believe the loudspeaker is the only musical instrument extant. It is easy enough to catalogue a set of OHP transparencies, but a good cataloguing system should make reference to other forms of AV aid which are relevant. This does not only require a considerable effort in cataloguing, it also requires a very detailed knowledge of the technology which is being dealt with; worthy of a Mastermind candidate.

Classification of material must be done with considerable co-operation from the users, even when all material forms part of one specialized subject. The form of the catalogue must be usable by all, which often rules out the computer as the only display method. Nevertheless, software such as *Sage Retrieve*, which can be run on any IBM-compatible computer, forms the basis of a very useful cataloguing system which can provide printed output for the benefit of those who (1) refuse to use the computer, (2) cannot use the computer (3) who need the information when the computer is locked away. Any database program used for cataloguing should be of the type called *relational*, so that it is possible to prepare and call up related topics easily.

A possible layout for data is illustrated in Figure 1.9, using a main heading under which the catalogue item is indexed. There is a brief description of material, and two sets of entries for other AV material on this subject, and for related headings under which relevant material can be found. Printed out this forms a useful loose-leaf catalogue entry and, when used with a computer, allows the user to browse over all the relevant material. Incidentally, some computer database systems allow for small sketches to be included in a record, though this requires a system with a large-scale backing memory, such as a hard-disk of 100M or more for a large AV department.

VISAID

Heading:	Steam locomotives	
Format:	Slide set + Tape	
Date created:	14th May 1990	

See also:

James Watt	OHP	Tape
Steam engines	OHP	
Railways	Slide	Tape
Steamships	Slide	Tape
Road engines	Slide	
Latent heat	OHP	Tape
Heat and work	Slide	

Figure 1.9 *Possible layout of a data entry system which allows cross references to other topics and forms of audio visual material*

Hints and tips

- Felt-tipped pens dry up quickly, and should be capped when not in use – some types cannot tolerate even a few minutes exposure. A few types can be sharpened when tips become blunt.
- Card mounts cost about £20 for a pack of 50 and allow valued transparencies to be kept in better condition over long periods.
- Fold-up portable types of OHP are easier to store, carry and set up than older rigid versions.
- Non-folding types of projectors are more easily moved if a transporter table is used.
- Magnifiers can be clipped to the lens of the OHP, allowing a 2.5 magnification of material, so fine detail can be displayed more easily.
- Where OHPs are used in a large variety of sizes of rooms, a set of different lens heads (standard, wide-angle and narrow-angle) can be useful.

- Many modern machines are fitted with *fringe eliminators*, optical systems which do not display coloured fringes on images.
- A useful feature is a two-position lamp brightness switch. If the lamp is always switched on in the dimmer setting, and used in this setting if feasible, its working life is greatly extended.
- Fresnel lenses can be obtained with a scratch-proof and anti-glare coating, both very desirable features.
- Quartz-halogen bulbs must *never* be touched by bare hands – always use rubber gloves when handling. If a bulb is touched it should be cleaned with surgical spirit (absolute alcohol) – never use methylated spirit.

2
The slide projector

The slide projector was at one time, along with a radio, the total extent of AV resources in schools. Even now, after all the developments in audio-visual technology over the past 30 years, the slide projector is still a very useful method of putting over some types of material, particularly when complete slide shows can be packaged and, if necessary, synchronized to a tape commentary. Advantages of slide use are (1) the compact nature of the projector and of slides themselves (2) the ease of making slides (3) the huge library of material available on slides (4) the ways in which slide shows can make use of effects such as fade-in and slow animation. Slide projectors are widely available to buy or to hire and the 2″ × 2″ (35 mm) slide format, with a picture area of 36 mm × 24 mm (Figure 2.1) is standardized throughout the world. The main drawback of slide use on the other hand, is that the brightness of the screen is comparatively low, requiring partial blackout in the viewing room. This can be overcome to some extent if back-projection can be used.

A recent development is of computer-controlled slide displays, allowing for the totally random selection of slides from the carrier of a carousel type of projector. Apple and IBM computers can be used with cables ready-made, or a cable can be supplied which can be connected to other computers if a suitable plug is available; otherwise a separate controller such as the *Kodak S-RA 2500* keyboard control can be used. At the other end of the technological scale, write-on slides are obtainable which allow the slide projector some of the versatility of the OHP, though requiring considerable skill in miniature writing techniques.

Virtually all modern projectors are remote-controlled, using a plug and socket connection to a remote control panel, and several models now feature auto-focus using the same type of infra-red system as still and video cameras. Infra-red remote control can also

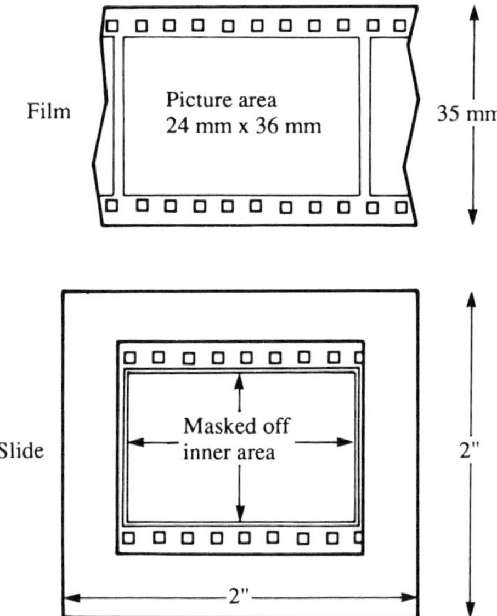

Figure 2.1 *Dimensions of 35 mm film and slide*

be used to avoid trailing cables. A point to note here is that infrared systems may interfere with each other and though the auto-focus of a slide projector is not generally a problem, there can be interaction of beams between other devices. You might, for example, find using a remote-controlled projector along with remote controlled video recorder and TV presents problems of interaction.

The old-time wooden pointer has now been replaced with a laser-beam device which is about the size of a TV remote control and which, in normal room lighting permits pointing at a distance of up to 20 m; further in dark conditions. Unlike a torch beam, the laser beam does not discernibly spread out as the distance is increased, so its pointing action is extremely effective. Further advantages are that it does not cast a shadow and is pocket-sized. Battery life is of the order of seven hours' continuous use, considerably more when the laser is flashed rather than as a steady beam.

Principles

Optical principles of the slide projector are much the same as those of the OHP, but without the 90° angle reflection of the beam. Light from a bulb, usually a low-voltage quartz-halogen type, is collimated into an even illumination for the slide, and the beam is then focused on to a screen as shown in Figure 2.2. Distances of slide to lens and lens to screen, along with screen width, are shown in Table 2.1 for a lens of 93 mm focal length (the usual lens supplied with slide projectors), and Table 2.2 shows how projection distance, focal length and picture width (from a 35 mm slide) are related.

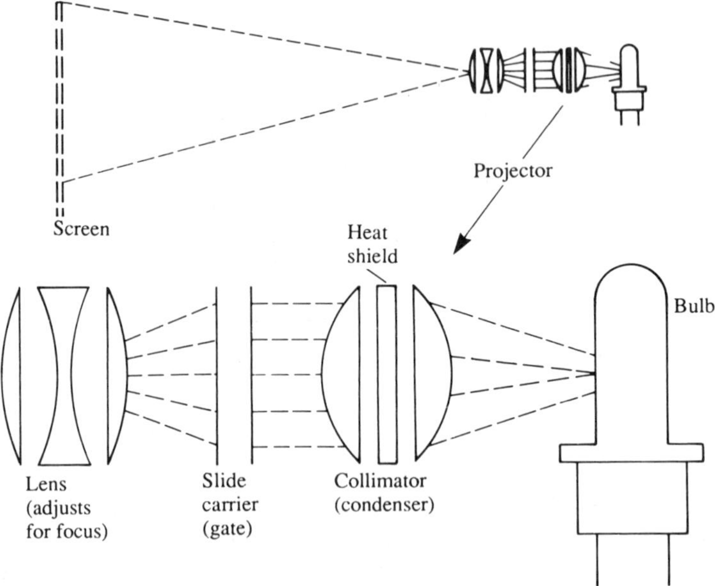

Figure 2.2   *Optical principles and light path of a slide projector*

Requirements for even illumination of the slide are much easier for a slide projector than for an overhead type since the slide is normally a 35 mm type, with usable dimensions of 36 mm × 24 mm. A few machines allow for the older 2.25″ slides to be used, but such slides are by now very uncommon. Other film formats such as 35 mm half-frame (as used on the *Yashica Samurai* cameras), and 126 size can be catered for by using conventional slide mounts with masking. Cooling in the area of the slide is very important because the light and heat from the bulb is concentrated onto such a small area.

Collimator or condenser lenses usually consist of several elements, with an infra-red absorbing glass between elements. The usual type is aspheric, meaning the curvature of the glass is not spherical, allowing correction to be made for inevitable optical

Table 2.1   Distance of lens to film and lens to screen for a lens of 93 mm focal length, along a screen picture width for a 24 by 36 mm slide

Screen to lens (m)	Lens to film (mm)	Picture width (m)
1.00	102.54	0.35
1.25	100.48	0.45
1.50	99.15	0.54
1.75	98.22	0.64
2.00	97.54	0.74
2.25	97.01	0.83
2.50	96.59	0.93
2.75	96.26	1.03
3.00	95.98	1.13
3.25	95.74	1.22
3.50	95.54	1.32
3.75	95.37	1.42
4.00	95.21	1.51
4.25	95.08	1.61
4.50	94.96	1.71
4.75	94.86	1.80
5.00	94.76	1.90
5.25	94.68	2.00
5.50	94.60	2.09
5.75	94.53	2.19
6.00	94.46	2.29
6.25	94.40	2.38
6.50	94.35	2.48
6.75	94.30	2.58
7.00	94.25	2.67
7.25	94.21	2.77
7.50	94.17	2.87
7.75	94.13	2.96
8.00	94.09	3.06
8.25	94.06	3.16
8.50	94.03	3.25
8.75	94.00	3.35
9.00	93.97	3.45
9.25	93.94	3.54
9.50	93.92	3.64
9.75	93.90	3.74
10.00	93.87	3.83

Table 2.2   Projection distances for a range of focal lengths

Lens focal length (mm)	Projection distance (m)											
	0.5	1.0	1.5	2.0	3.0	4.0	5.0	6.0	7.0	8.0	9.0	10
26	0.65	1.30	1.96	2.60	4.00	5.20	*	*	*	*	*	*
36	0.50	1.00	1.50	2.00	3.00	4.00	5.00	6.00	*	*	*	*
45	0.43	0.82	1.20	1.58	2.30	3.10	3.90	4.60	6.20	*	*	*
51	0.35	0.69	1.04	1.38	2.10	2.80	3.50	4.10	5.50	*	*	*
55	0.31	0.62	0.94	1.24	1.90	2.50	3.10	3.80	5.00	6.20	*	*
60	0.29	0.58	0.88	1.17	1.60	2.20	2.80	3.30	4.40	5.60	*	*
85	*	0.40	0.61	0.82	1.30	1.70	2.10	2.50	3.30	4.10	5.00	5.90
93	*	0.35	0.55	0.74	1.10	1.50	1.90	2.20	3.00	3.70	4.50	5.30
100	*	0.32	0.50	0.68	1.00	1.40	1.70	2.00	2.70	3.40	4.10	4.80
135	*	0.24	0.36	0.50	0.80	1.00	1.30	1.50	2.10	2.60	3.10	3.60
150	*	*	0.31	0.43	0.60	0.90	1.10	1.40	1.80	2.30	2.80	3.20
200	*	*	0.20	0.30	0.50	0.70	0.90	1.00	1.40	1.80	2.10	2.50
253	*	*	*	*	0.40	0.50	0.70	0.80	1.10	1.40	1.70	1.90

defects if spherical glass curvature were used. Kodak provide different collimators for use with the very long-focus projector lenses (150 mm and 180 mm). Use of infra-red absorbers greatly cuts down the amount of radiated heat reaching the slide, but not always to an extent sufficient to avoid *popping* (see later). The main projection lens is usually an optically simple type compared with a camera lens and, where a projector is to be used in a variety of different rooms, it is usually necessary to carry a selection of lenses to cater for different viewing distances. Tables 2.3 and 2.4 give details of ranges of lenses. For many applications, one normal and one wide-angle type are often sufficient. If back-projection (Figure 2.3) is used, a wide-angle lens is essential because distance between projector and screen usually has to be short. In addition, when straight (as distinct from mirror-path) back-projection is being used, slides have to be placed in the carrier with left and right reversal, not simply upside down (Figure 2.4).

Brightness of projected images is important because it determines the usefulness of a slide projector in rooms which cannot be thoroughly darkened. A back-projection system is often the only solution to viewing slides in normal lighting, but when some reduction in lighting can be achieved, modern high-brightness forward-projection screens can be suitable. The snag is that such

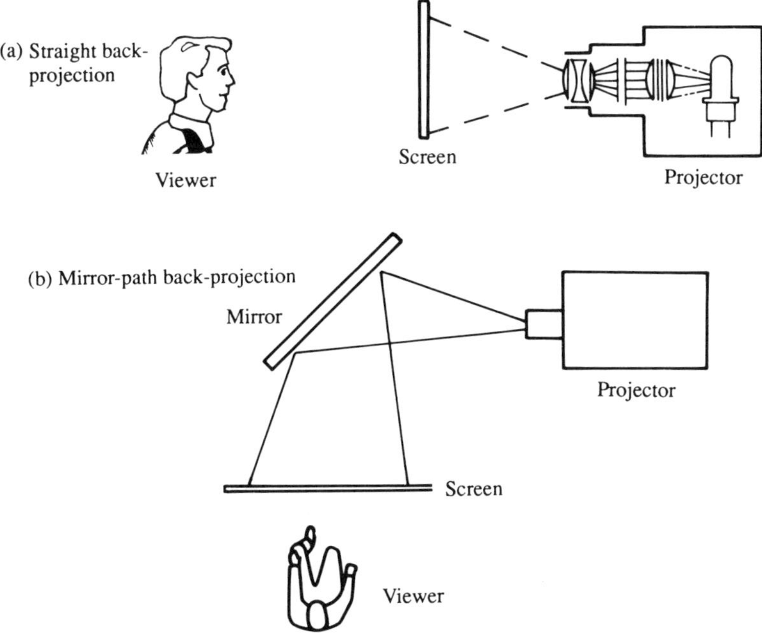

Figure 2.3 *Using back-projection with a wide-angle lens so that the distance from projector to screen is short*

screens, which rely for their brightness on being coated with tiny glass beads, provide an image which is bright only for viewers at a narrow angle to the screen-projector axis, as illustrated Figure 2.5. Outside this angle brightness falls off considerably, so such screens are better suited to long narrow rooms than to short wide rooms. The older type of ribbed or matte-surfaced screen is better suited to wide angle viewing, but the image is considerably dimmer.

Where a large viewing audience must see the screen, the usual arrangement of having the screen at about head height is not suitable, and an elevated screen must be used. This brings the same problem of keystone distortion as was discussed for over-head projectors and the same solution, of tilting the screen, applies. Most screens can be tilted forward by up to about 15°, which is usually enough to make the amount of keystone distortion acceptable. Another approach to keystone correction is available for the Kodak *Carousel* range of projectors. By using the *Kodak RETINAR S-AV 2000 PC* lenses, keystoning can be eliminated even for wide-angle projection by altering the positions of the internal components of the lens – the shift mechanism can be locked when a

Table 2.3   Kodak Retinar S-AV 1000 lenses

Catalogue number	Focal length (m)	Length (mm)	Diameter (mm)	Weight (gm)
701 8450	55	125	73	234
701 8468	85	125	73	108
701 9116	100	125	73	108
701 8476	150	146	73	206
701 8484	180	190	73	205
701 3801	Zoom lens, 75 to 120 mm focal length			

Table 2.4   Kodak Retinar S-AV 2000 lenses

Catalogue number	Nominal focal length (mm)	Specified focal length (mm)	Aperture	Length (mm)	Diameter (mm)	Weight (gm)
701 8492	26	26.4±0.3	f/2.8	151	73	721
701 8500	36	35.4±0.3	f/2.8	125	78	457
701 8518	51	50.6±0.5	f/2.8	125	78	389
701 8526	93	92.8±0.5	f/2.5	125	78	320
701 9090	135	136.0±0.7	f/2.8	125	78	437
701 8534	150	149.9±0.7	f/2.8	146	73	605
701 9108	200	200.5±1.0	f/3.5	164	73	618
701 8542	253	252.5±1.0	f/4	214	73	870

suitable setting has been reached. This range of lenses can be fitted to older *Carousel* projectors as well as to current models.

Slide handling mechanisms

Slide projectors in the ages BC (before carousel) were generally manufactured by Leitz or Aldis and used hand loading and slide-changing. Where such projectors still exist, they should be carefully maintained for two reasons: (1) their value is steadily increasing (particularly for pre–1939 Leitz types) and (2) they often offer the only way of showing 2.25″ slides. Since the introduction of the Kodak *Carousel* in 1964, the older type of manually-operated slide projector has become almost obsolete.

Magazine-operated projectors are either of the circular-magazine (like the *Carousel*) or straight-magazine type. The obvious

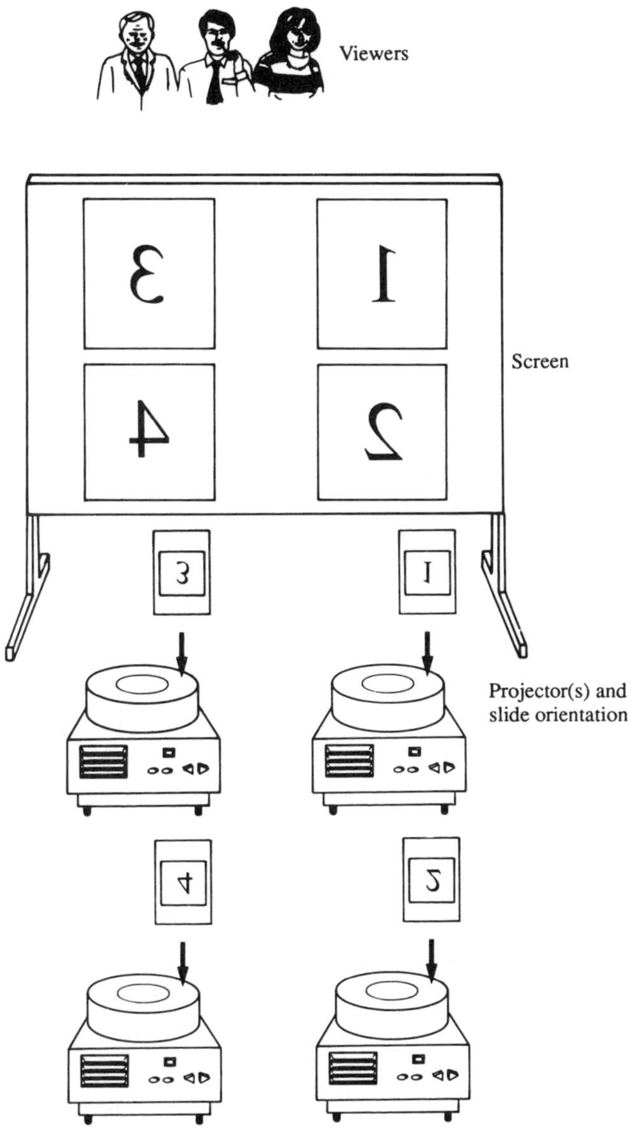

Figure 2.4 *When back-projection uses the straight light-path, slides are put in with left-right reversal as well as upside down*

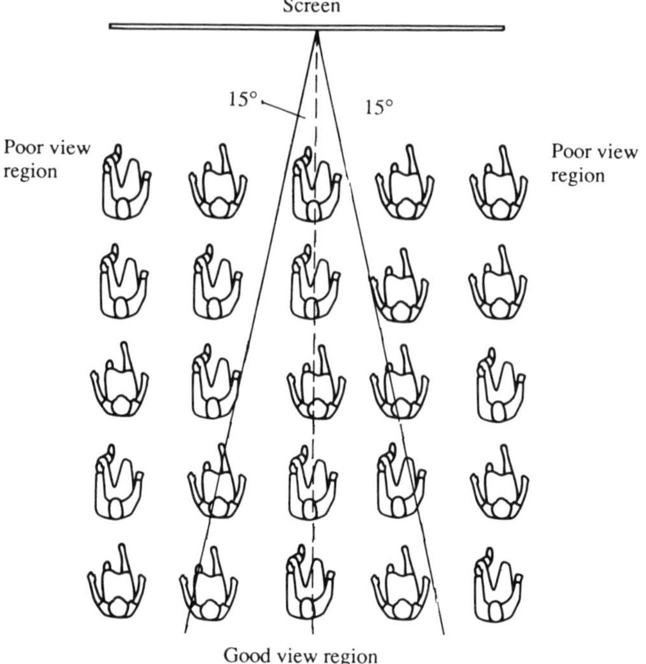

Figure 2.5 *The narrow viewing angle of a glass-beaded screen. Outside this region, brightness of projected images is considerably lower*

advantage of the circular-magazine type is that a slide show can be completely repeated as often as is needed, and is ready to repeat once the set of slides has been shown. The straight-magazine type is more compact and the magazines are easier to store, but the circular *Carousel* type is employed to a much greater extent both in education and in business, with the straight-magazine type used mainly for domestic uses. There are, however, excellent straight-magazine projectors available from Leitz and from Rollei, and the Rollei models feature the ability to fade from one slide to another by using twin lens projection on a single machine, an action that normally requires two projectors. The most common capacity for magazines is 80 slides, which is more than enough for most illustrated talks.

Gate mechanisms for slide projectors of the straight-magazine type use a solenoid or a motor for their action, and are easily jammed. The most common cause of jamming is the use of sticky labels on slides. These labels almost invariably detach themselves when heated, and seem to have the remarkable power to attach themselves to the gate mechanism with considerably better adhesion than they had to the slide. Even when such a label is removed it will leave a sticky patch on the gate which will cause another slide to stick at some other time.

The mechanics of the gate are by no means simple. Taking the *Carousel* as an example, the slide tray has a floor which contains one slot, slot 0, wide enough to allow a slide to fall through (Figure 2.6). This slide position is empty when the tray is being transported, but when the magazine is put into place and the projector started, pressing the slide-change button turns the tray, so the first slide is over the large slot and falls into the gate for projection. When the change button is pressed again, a shutter is placed in front of the collimator to blank off light to the slide, and the slide which is in the gate is pushed up by a lever back into the carousel or slide-tray. The carousel or tray then advances to the next position, and the slide in this position falls into the gate when the ejection lever drops again.

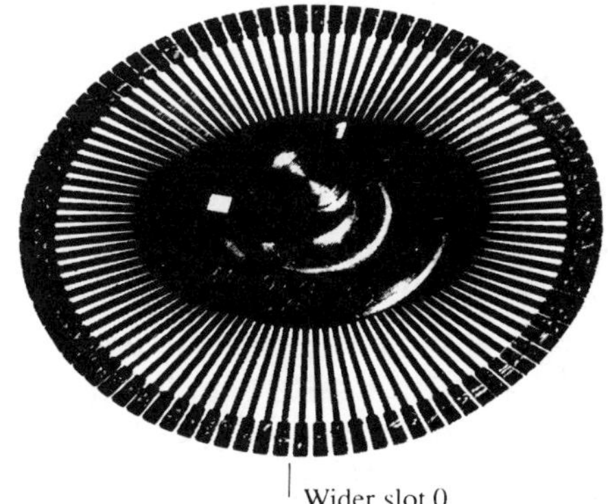

Figure 2.6   *Carousel tray showing the large slot through which each slide falls in turn to be projected*

The advantage of the *Carousel* system is that gravity (a most reliable mechanism) is used to move each slide into the gate; the slide simply falls through the wide slot in the floor of the tray. The mechanical arm lever action is used only for replacing the slide, and this only has to move the slide to just above the floor level of the tray: movement of the tray itself then takes the slide beyond the wide slot as the ejection arm drops again. In contrast, straight-tray machines use a side mounted tray and both insertion and removal of slides is done by a moving arm. This mechanism (certainly in my experience) seems to be more troublesome, particularly when slides are in cardboard mounts, and it is not unusual to find the arm missing the slide altogether and subsequently jamming.

Another common problem is that the slide tilts when pushed so it jams the entry to the gate.

The gate mechanism of the *Carousel* needs little maintenance, a point which has ensured success of this type of projector. At most, the floor of each tray needs to be removed, cleaned and sprayed sparingly with silicone lubricant after several hundred hours of use.

Slide-popping is a problem which has been present since the practice of encasing slides in glass ceased. Older 2.25″ slides were almost always mounted by pressing them between two flat glass squares, but this practice, beside being costly, caused two major problems. One was the light-interference pattern called Newton's rings which is caused when two almost flat transparent materials are pressed together. This pattern of distorted circles is almost unavoidable when glass slide-carriers are used and can be very obtrusive. In addition, because the glass slide carriers insulate the film, it is easy to overheat the material, causing colours to fade rapidly and black areas of monochrome film to burn.

Slides are now almost always 35 mm, and use plastic or card holders which retain the film around the edges, with no glass pressed against the film. The plastic types are of the click-fit variety, in which the film is put into place and the two halves of the carrier pressed together until they click, locking the halves. Card type use pressure-sensitive adhesive on each half, which stick when contact is made. In either case, film is exposed directly rather than through glass. Though this makes slides more fragile and more susceptible to fingermarking, it greatly improves cooling of the film, so the effective life of the slide is better. Many projectors can feed the plastic carrier type of slides more reliably than card-mounted types, and there is the further advantage of not using any form of adhesive, a useful point if a film has to be removed from a damaged carrier.

Use of 'open' slide carriers of this type, however, permits popping. When a slide is assembled, the film has a natural curvature, though it may be stretched flat in the casing. Under the heat of projection, the curvature of the film reverses as the side that carries the dyes of the image absorbs heat. This is the popping process, so called because the sudden reversal of curvature often causes a popping sound which can be heard if the projector fan is quiet. Popping is particularly prevalent on modern projectors with quartz-halogen lamps because these provide a more concentrated illumination than older types of projection lamps.

The effect of popping is to alter the focus requirements. If the central area of a slide is in focus before a slide pops, it is almost certainly out of focus after popping. Further, when a slide surface is curved, as it is after popping, it is impossible for both centre and edges to be in perfect focus together – you can focus for the centre *or* the edges, but not both. This need not cause a great amount of trouble because if the amount of popping is small, a compromise focus setting is usually found which provides a reasonably clear picture. The alternative is more radical: to use a lens which focuses from a curved slide, such as the Kodak *Curved-field* lenses. The extent of popping is considerably less with modern films and with holders which retain film more firmly. Projectors which feature auto-focus self-adjust when a slide pops.

Tape-slide synchronization

Magazine projectors advance to another slide when an electrical connection is made between two points in the projector. External sockets allow this connection to be made by remote-control units connected by cable, and also permits synchronization from tape. It is important to note, however, simply connecting an ordinary cassette recorder to a projector does not have any synchronizing effect.

Almost universally, tape-slide synchronizing methods use a recorded commentary on one track of a twin-track tape and use the other track exclusively for synchronizing signals, a brief burst of a tone at 150 Hz (low note) or 1000 Hz (high note). Use of two notes allows one tape to synchronize two projectors independently. The audio commentary track is replayed through loudspeakers in the normal way, but signals from the synchronizing track are amplified and used to operate an electronic switch connected to the remote-control socket of the projector, allowing a slide change to be made when the correct tone occurs on the synchronizing track.

The two main methods of making synchronizing tapes are the use of a specialized recorder (such as the Kodak *S-AV Cassette Recorder 200* or the ITT *SL537AV*) or by using a tape-slide synchronizer along with a conventional tape or cassette recorder. Use of a dedicated recorder allows tapes to be recorded in two-track format which yields more reliable synchronizing actions, but separate synchronizing units are popular where an AV department possesses many ordinary cassette recorders. In addition, it is possible for a technician with electronics skills to construct slide synchronizing units either from a circuit diagram or from one of the kits currently available (see the advertisements of Maplin Electronics, for example). The unit is connected to the input/output 5-pin socket of the cassette recorder and to the remote socket of the projector (see later).

User guide

Users of slide projectors need to be aware of the need to set up the screen and projector correctly, including lens selection if room requirements vary, and methods of loading slide trays and using the remote control system. In addition, users need to be able to specify slides used, either by hiring a slide set, choosing slides from a library collection, or taking photographs using slide film (on field trips, for example).

The first point to attend to is the extent to which a room can be darkened. Purpose-built lecture theatres present no problems here, likewise conference rooms of hotels, but teachers working in the sort of cl— —ms which were originally part of an asylum until being jud—   such use tend to be at a disadvantage. The worst   ngly enough, are those in purpose-built schools wi...   won architectural awards and which are magnificent buildings for architects' offices but quite unfit as schools. Windows are usually large, south-facing, and with such narrow frames that no blind can ever prevent shafts of sunlight coming into the room, assuming that any of the blinds can be made to descend. The converted Bedlam, by contrast, often has high slits of windows with wide frames and blinds that work.

The difficulty of finding a room suitable for projector use in a modern building often makes it necessary to book a place in a specifically appointed AV room, and to ensure the class goes there on the appointed day and time. A problem often arising in this respect is that of permanent bookings (the art department running its slide show on History of Art every day between 10 and 11), which makes equal sharing of such facilities very difficult. Even when a room can be adequately darkened ventilation may be inadequate so that the use of slide projectors in summer term is almost out of the question. The design of schools since the 1960s has done more to stamp out the use of visual aids than the usual constraints of falling grants and rising costs.

Another problem is screen placement. Where a room has a permanent whiteboard, this is usually expected to double as a screen, but may be badly placed or of material so glossy it dazzles the viewer rather than displays an image. Screens arranged as roller blinds permanently mounted above a blackboard are often well-placed but only too often easily damaged, so the usual expedient necessary is the erection of a portable screen, which can be packed away safely after the slide show. At least use of a portable screen permits a choice of screen position, and as long as the screen can be elevated and tilted, a good picture can be obtained (see section on keystone distortion, Chapter 1).

The projector itself has to be set where it is unobtrusive to the viewers, easily manipulated and optically suitable – ideally on a high stand at the back of the room. Use of a high stand allows a fairly high screen position with the minimum of keystone distortion, and in such a position the projector is unlikely to be in the way of any viewers. In a specialized lecture theatre, of course, the projector can be situated in a separate projection room, remote controlled from the lecturer's bench. Fine adjustment of projector position is controlled by the image height and levelling adjustments on the projector itself.

Use of magazine-fed slide projectors allows almost as satisfactory facilities to be available in any room providing three conditions are met. One is that the remote-control lead should be long enough, assuming infra-red remote-control is not in use. Many slide projectors still rely on cable-attached remote-controls so that too often, it is possible to site a projector in an ideal place at the far end of a room, only to find it must be controlled from a far less ideal position about half-way down the room. Even with the cable type of remote control there is no excuse for this because it is simple to make cable extensions in several lengths. Use of a laser type of pointer also obviates the need for the user to be near to the screen to point out details of interest in the displayed picture.

The second condition to be met is that the projector should have a lens whose focal length is matched to the length of the room. Small rooms require a wide angle short-focus lens, so a picture of reasonable width can be projected. Keeping the same lens in place when a longer room is used means the projector needs to be sited half-way down the room, or that a very much larger screen is used, neither of which is an ideal way of working. Projector lenses are modestly priced, and there is no reason why a variety of lenses should not be available. It is not usually necessary to have a complete set of lenses for each projector in use, only a few spare lenses of long focus for long rooms and short focus for short rooms.

Given good room conditions and siting of projector and screen, the third and final proviso is in the hands of the user: the slides themselves. These must have been loaded into the carousel and checked before the presentation unless a carousel can be left packed with slides for a specific presentation. Most users are obliged to keep slides in small boxes and to load up the carousel just before using the projector. This makes it important to sort out the order of slides in advance, and to ensure they are in the correct position before the presentation starts. Nothing creates more distraction than having the first slide of a set the wrong way round, or of the wrong subject.

When the slide show is of a set permanently in the hands of the user there should be no difficulty about numbering the slides so they can be quickly loaded into the carousel. Numbering should be done using an Indian-ink pen rather than a felt-tipped pen or labels. Ink from most varieties of felt-tipped pens smears badly when the slide is handled, and labels peel off in the heat of the projector gate and cause the slide to stick. When a set of slides has been borrowed from an educational institute or from an AV library, however, numbering cannot be done, and the slides have to be checked and put immediately into the carousel. If the AV centre or workroom has a slide viewer (no matter how simple, even an illuminated panel is useful) this task is made considerably easier. The *Carousel* type of projector can also be used as a slide checker by removing magazine and tray and dropping a slide into the gate aperture. The slide can be ejected by using the slide-change switch. Another option is to use the straight Kodak stack-loader to look at a set of up to 40 slides, though this does not allow the slide selection to be made backwards – once a slide has been viewed you cannot return to it from the next slide on. Once slides are in place on a full-sized carousel magazine it is useful to view them on a spare projector, using a back-projection or side projection set-up such as discussed earlier. This type of slide-checker can be set up using an old projector or one always kept as a spare. In the absence of the facility to check the slides in this way, you have to rely on the older strategy of crossing your fingers.

Good preparation is particularly important when a matching slide-tape show is to be used, because any mismatch here can be hilarious and can take a considerable amount of effort to correct. A live speaker need not be thrown by a misplaced slide, but there is simply no easy way of ensuring tape and slides can recover synchronization after losing it. The nightmare of anyone who depends on such a presentation is of spilling the slides just before starting, which underlines the importance of keeping the magazine covered until it can be loaded into the projector.

Preparing a synchronized tape-slide show should start with a rough draft of the topics and order of slides. Slides can then be assembled into the magazine, and more complete notes on the accompanying talk prepared while viewing the slides. Take time over this stage because, as they say, mistakes are more difficult to rectify afterwards. Once all slides for the presentation are selected and put into the correct order, the talk can be recorded. This is done using the specialized synchronizing cassette recorder or with a tape-slide synchronizing unit connected to both recorder and projector. Make sure the leader of the tape has been wound on (see Chapter 4) and the tape is placed the correct way round in the recorder (starting the tape rather than ending it). Check that the microphone is correctly connecting and is recording, and that the synchronizing pulses are recording correctly – these checks are easier to make (without recording anything) on a specialized machine.

With the first slide showing, start the recorder and begin the talk. If you cannot maintain continuity, stop the recorder at once while you gather your thoughts. At each point where a slide change is needed press the synchronizing button, so a tone is recorded on the spare track. Some synchronizers provide another tone, recorded at the end of the talk, to switch up room lights, or a portable light near the recorder.

Once the tape is complete, rewind it and run the magazine of the slide projector round to the starting point. With the first slide in the gate, start the tape and check the correct slide changes are being made. Make changes only if absolutely necessary, because any major change involves recording commentary and synchronizing tones again from the point where an error is found.

Material for slides makes this one of the easiest visual aids to provide, because anything which can be photographed can provide a slide, and anyone who can handle a camera can make slides. Of this number about 10% can make consistently good slides using older types of cameras, but the increasing use of fully automatic cameras now makes it possible for virtually any user to create slides of viewable quality. There are still specialized actions such as making slides from a TV image, from a microscope, or from book illustrations which are properly within the domain of the AV technician, but slides of machinery, animals, buildings, and so on are now within the capability of any user. Another useful point is that colour slide processing has always been considerably simpler than any form of printing, so slide film can be processed at low cost and with a minimum of equipment, see Chapter 7.

A point often overlooked is that ordinary black and white film can be processed into slides (reversal processing) and such slides can be of excellent quality, ideal for displaying fine detail. Processing is even simpler than that for colour slides and requires only the most basic equipment of a photographic changing bag, a developing tank and a place in which to work. Further details of the process are noted in Chapter 7. Another possibility is to use a 35 mm camera with a slide-copier attachment to photograph existing *negatives* which have been used for prints. When the film is processed, the resulting frames can be used as slides, since the colours will be reversed from the original negative form (or black and white interchanged). This is more satisfactory than copying prints using slide film, particularly as developing colour print film is so inexpensive.

When the time comes to start the slide show, it is always an advantage to make use of any high/low lamp brightness control fitted. On the *Carousel* type of projector, this control, if fitted, depends on the setting of the int/ext lamp switch, which should be set to *Int*, allowing internal control. Use of the low brightness setting at switch-on has a number of advantages, not least of which is that lamp life is considerably improved by switching on in this setting, since it reduces the surge of current through the cold filament of the bulb, the usual cause of bulb failure. The extent of this improvement can be remarkable – one projector in my hands used the same bulb for ten years because it had been adapted to automatic low-brightness switching using an electronic system which kept the brightness low for the first ten seconds after switching on the bulb. Further, the lower setting may provide all the brightness you require, making it unnecessary to switch to a higher setting, thus extending the life of the bulb even more.

Technical maintenance

Regular maintenance of slide projectors of circular or straight magazine types amount to little more than cleaning and bulb replacement, and there should be no need for lubrication of any of the moving parts of the projector itself in its working life. The main factor that can significantly shorten the life of a slide projector is operation at consistently high temperatures, and this applies mainly to projectors operated by remote-control in soundproof booths or smaller enclosures. Best test for operating conditions is to measure air temperature from the exhaust aperture at the rear or side of the projector. This should never exceed 71°C (160°F) after the projector has thoroughly warmed up, and temperature should be taken by allowing a thermometer to remain in the centre of the exhaust grille for at least two minutes.

If the temperature of air *entering* the projector is too high (above 30°C) the thermal fuses in the projector are likely to melt, preventing the equipment from being used. Thermal fuse spares should be held by the AV department. Never attempt to make any temporary connection to replace a thermal fuse. Quite apart from being foolhardy, this would render you liable to blame in the event of a fire arising from overheating equipment. Whether the projector is used in a confined space or not, always check vents are unobstructed before a projector is passed out for use. *Carousel* projectors have a small air vent which *heats* the slides themselves, and this must not be neglected – it is visible when the tray is removed. The purpose of this vent is to ensure that slides are at about projector temperature when they reach the gate, so any popping is completed by this time and there is no need to keep altering the focus setting.

Normal maintenance starts on a projector which has been allowed to cool by disconnecting the projector from the mains supply, and removing the remote control and the slide tray. Projectors in intensive use should be fully serviced at 500-hour running intervals, and if they are used under unusually dusty conditions (such as exhibitions) servicing should be more frequent. The main projection lens can be removed and cleaned with a lens cloth, using only a very light pressure on the surface. The main source of trouble is fingerprints, and the lens should be checked after polishing to ensure it is free of such marks. The condenser assembly should then be checked, including the infra-red absorbing layer. Infra-red absorbing glass is liable to shattering, rather like the toughened glass windscreens fitted at one time to cars, and goggles should be worn while handling it at all times. When laid aside awaiting re-assembly the infra-red filter glass should be kept covered so if it should shatter the small pieces are confined.

A complete service involving taking the projector apart should be done only if the full technical manual is available, or if you have experience of this work. Some projectors require a full service to be carried out at a service centre while the machine is within its guarantee period, but there is no objection to the experienced AV technician carrying out a full service on a machine that has been in use for many years. Care is needed with lubrication – lubricate sparingly only where indicated in the manual, and using only the specified oil. Never assume a drop of 20–50 will do in place of the specialized lubricant the manufacturer specifies.

Routine maintenance other than cleaning generally involves bulb changes. Older projectors used in the UK generally use 24 V 150 W quartz-halogen bulbs, but the later types use 82 V 250 W bulbs which are also specified in projectors of US or Canadian origin using the *Ektagraphic* trademark. The *S-AV 2055* projector uses a 36 V 400 W bulb, which is not easy to obtain if you need a replacement in a hurry. Where a projector uses a dual-bulb system which allows a bulb to be mechanically switched over to a spare without removing the faulty bulb, both bulbs should be tested. A user may have switched bulbs and not have reported a bulb failure or may have switched back to the faulty bulb after using the projector.

If a bulb has been run just before servicing, it must be given time to cool before any attempt is made to change it, and the new bulb should be kept in its packaging. Usual rules for working with quartz-halogen bulbs apply: the quartz envelope of the bulb must not be touched by hand because fingerprints will bake to carbon, reducing light levels, causing overheating and reducing the life of the bulb. Keep a pair of rubber or plastic gloves handy for making bulb changes unless the bulb is one of the type supplied with a protective handling sleeve.

On dual-bulb projectors the switch operating the quick bulb change usually has a third position for bulb removal, allowing the lamphouse door to be opened and the bulb holder to be extracted. The bulb holder on *Carousel* machines is hinged and can be swung out when the lamphouse door is opened. Always use a glove to grip the faulty bulb and pull it from its socket, then replace it, handling the new bulb by its protective sleeve or with a glove. If by chance a bulb is touched leaving a fingerprint on the quartz, wash the bulb with cotton-wool soaked in absolute alcohol (methylated spirit is a poor second choice, as it usually leaves a film of material over the bulb). Follow the detailed instructions for your particular model of projector after plugging in the new bulb.

In particular, it will probably be necessary to centre the new bulb, using two adjustments, one of which moves the bulb holder vertically while the other moves the holder horizontally. The extent of the adjustment has to be monitored by watching the light spot created from a test slide, ensuring no shadowing occurs at the edges due to mis-centring of a bulb. Check that centring is correct for both bulbs of a pair if the projector is of the twin-bulb type. At a bulb change, check also the thermal safety switch to ensure that it is clean.

Rather more regular attention is needed for slide trays, particularly as there can be many slide trays for each projector. Slide trays may be specific to projectors, though this is usually only a problem if some old models of projectors or trays are in use. While projectors are usually taken to and from their room of use by the technician and spend little time in the hands of the user, slide trays may be out of the technicians reach for months on end, sometimes until a fault develops. Slide trays should be kept clean, and if sticky action is reported, lubricated very lightly with silicone spray oil.

At times, a projector may be returned with a slide jammed in the mechanism and bent. It should still be possible to remove the slide tray (by holding the latch aside while lifting the tray) and once the tray has been removed the slide can be taken out and transferred to another holder. In this respect, *Carousel* projectors are very much easier to deal with than the straight-tray variety, some of which need to be almost completely dismantled to remove a jammed slide.

At times, the technician is called on to recover slides soaked in water, coffee, lemonade, and so on. This calls for some co-operation from the users, because the first important point about water damage is that it need not occur if the slides are kept wet – damage is done as slides dry. As soon as possible after wetting, slides should be immersed in water, and if possible, formaldehyde preservative should be added to the extent of 0.5% formaldehyde in water (corresponding to 15 ml of commercial formalin solution per litre). Formaldehyde prevents bacterial growth and the swelling of the emulsion.

Slides preserved in this way can be dismantled when convenient, and washed in clean water. They are then rinsed in a solution of wetting agent such as Kodak *Photo-flo* and subsequently rinsed in *Kodak Stabilizer* solution, and are finally dried slowly in a dust-free space. The slides are then placed into new mounts ready for use again. In general, this type of treatment is also effective in removing marks caused by bacterial growth, mainly due to storage in warm moist conditions.

Storage of slides should be considered carefully. Never store slides in the same area used for chemical storage, because several chemical vapours affect slides noticeably. Most notorious of these is p-dichlorobenzene, as used in moth-balls, because vapour of this chemical causes crystals to form on film and damages the adhesives used in card mounts. A danger here is that moth-balls themselves are not usually considered a hazardous chemical – cameras and film must also be kept clear of moth-balls. More recognizable hazards are gases which have reducing or oxidizing action on the dyes of the film – gases such as nitrous oxide, hydrogen sulphide and sulphur dioxide. Dust particles, particularly of alkalis and of photographic chemicals (including hypo) will also cause the dyes to fade.

Where the technician is expected to construct equipment as well as maintain it, one very common requirement is for remote-control and interfacing units. These require electrical/electronics skills, and for the benefit of readers who might need to make up units, connections for the two types of sockets are illustrated in Figure 2.7 (the 12-pin type of socket) and Figure 2.8 (the 6-pin DIN type of socket). There is a US publication, *The Source Book*, available from the Motion Picture and Audio-visual Markets Division, Rochester, N.Y. 14650 or from Kodak UK, which is packed with useful hints for construction of equipment, wiring of projectors with controllers and other hints and tips. Any technician involved in the use of *Carousel* projectors should have access to a copy.

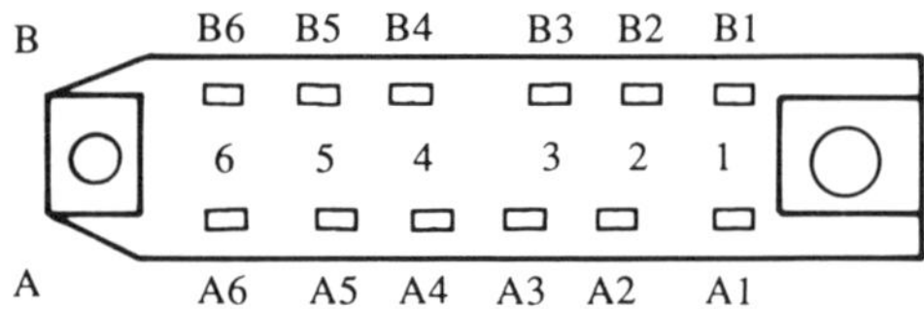

A1–A4 24V AC (750 mA max)
B4–A5 20V DC pulse (750 mA max)
A3–A4 External control
B5–B6 Slide tray zero reset
A6– Zero reset switch in gate
B3–B4 Forward slide change
B2–B4 Reverse slide change
B1–B4 Snap change

Figure 2.7   *The 12-pin type of socket for interfacing units, along with a list of connections*

Where extensive use is made of slide-tape shows, it is useful to copy tapes so a master is held in stock, and to remove the record-tab from each tape used for slide synchronizing. This avoids the considerable waste of time if such a tape is wiped or damaged.

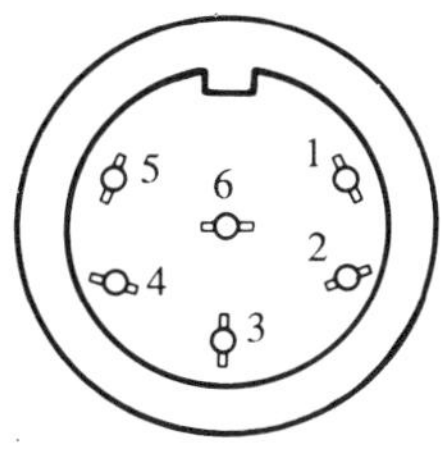

Connections

2 – 3 forward side change
1 – 3 reverse slide change
6 – 3 20 V dc 750 mA max

Figure 2.8   *The 6-pin DIN type of connector commonly fitted on Carousel models for remote-control purposes*

Slides used in any synchronized show *must* be numbered in their showing order in case (1) they are ever spilled from the magazine, or (2) a shortage of magazines means slides are to be kept in another case between showings. Synchronizing units present few problems; the tones used will record and play satisfactorily even if the tapeheads badly require servicing, and the main items to check are the fit of plugs and sockets.

Hints and tips

- Where the machine uses a receptacle for its mains cable, always pull the cable completely out when the projector is in use, even if the full length of the cable is not being used. This avoids damage of the cable due to overheating.
- Kodak projectors for the European market are made in Stuttgart and identified by the *S-AV* prefix to the type number. Spares, including lenses, are not generally interchangeable with the US/Canada models. A large number of projectors of the US type (with 240 V transformers) become available from time to time on the UK market, usually from USAF bases, and though the prices are attractive, they should not be bought even as machines for emergency use.
- *Carousel* tray-bands are a convenient way of labelling a magazine permanently used for a particular set of slides.
- A *Carousel* tray can be removed in an emergency (such as a jammed slide) by holding the magazine latch (at the hub of the magazine) aside and lifting the magazine out.

- Slide duplicating holders, with suitable adapter lenses, are available for most modern cameras. These allow copies to be made of important slides so that the originals can be kept in a cool dark place. Black-and-white copies of colour slides can be made, and are sometimes useful.
- Similar copying holders are available to allow slides to be made from photographs or book illustrations (note: book illustrations may be the subject of copyright).
- If cotton or nylon gloves are always worn when handling slides, the incidence of fingermarking is eliminated.
- Try to remove fingermarks from slides before they are projected, as there is a danger of baking the marks on to the slide.
- If tape for a tape-slide show has been repaired, check the join does not cause false triggering of the synchronizing device.
- Always keep slides and tape for a synchronized show together. Controlling your projector with a recording of *Eine Kleine Nachtmusik* is certainly entertaining, but is unlikely to be what you intended.

3

16 mm film projectors

Video has greatly reduced the dominance of 16 mm film as a medium for showing movie material, to the extent that information on movie film projectors is now hard to come by. Nevertheless, until LCD screens for OHP use with video (see Chapter 1) are widely available, movie film is the only medium available for working with large screens, and cost is very much lower than for video equipment with a reasonable screen size. Paradoxically, the exact opposite is true when you create films – the cost of film and processing alone for a 16 mm movie camera is several pounds per minute as opposed to a cost of a couple of pounds for as much as three hours of video. The cost of a video camera is now comparable to the cost of even a budget-range 16 mm cine-camera, and the cost of a video-cassette recorder is now comparable to the cost of a 16 mm projector.

There is still an excellent range of visual material available on film, often free or at very low hire charges, while the requirement for almost complete darkness for showing film is an aid to concentration. Movie projectors are of an old and familiar technology, easily maintained and rebuilt, unlike video equipment, and also reasonably portable and sturdy. Though some of the films that go the educational rounds are almost laughably old (dating to 1948 or earlier), there are a few gems seen as something new and unexpected by each audience. A select few are so outstandingly good it is worth retaining a movie projector for these alone.

Principles

Principles of movie projection are based on the older technology of slide projection superimposed on the work of Muybridge and others on animation. The patents of Le Prince show what amounts to a sequential set of fast-operated slide projectors, using 16 lenses, and work by Friese-Green and Edison later fixed the design of the single-lens machine using celluloid film. Celluloid itself (cellulose nitrate) is terrifyingly flammable, virtually a high-explosive, and fire risks involved in using celluloid led to its replacement by cellulose acetate safety film and in later years by various polyamide plastics. Old film on celluloid deteriorates continually due to the instability of celluloid and should be copied as soon as possible.

The basis of the illusion of moving pictures is the action of the eye, forming an image by a chemical action in turn stimulating electrical impulses in nerves which lead to the brain. Chemical actions have a definite time delay built into them of the order of $1/12$ second, so when a picture is flashed on and off at a rate of more than 12 times per second the eye sees a continuous picture, though with some flickering if the picture is bright. If the rate of switching on and off is increased flickering becomes unobtrusive and, for pictures of comparatively low contrast, is not noticeable if the rate is 24 pictures, or frames, per second.

Television pictures are considerably brighter than movie projected pictures, and this point was realized by the pioneers of the TV system we use today, the team led by Isaac Schoenberg at EMI in 1934–38. They decided on a switching rate of 50 per second, but because this would have required unacceptable high transmission rates settled on the compromise called interlace, in which half of the lines of a picture are shown in $1/50$ second and the other half in the next $1/50$ second. This gives a switching rate of 50 per second, but with the full picture being shown only 25 times per second. A similar dodge is used in movie projectors in the form of a multibladed shutter (see later).

In the movie projector, the main problem is and always has been to produce sufficient illumination for the film. As the film is moving, problems of overheating the film are less serious compared with, say, slide projectors, but frame size is small and so the enlargement that is used (ratio of screen size to frame size) is considerable. Inevitably, high enlargement means low brightness, so movie films have to be viewed in as near to complete black-out conditions as can be achieved.

The film cannot simply be wound continually through the gate of the projector, because the eye would never see an image in such conditions. Instead, each frame of the film must be held stationary in the gate for long enough to allow the eye to respond. Light must then be shut off, the film moved rapidly on to the next frame, then light restored so the next frame can be viewed. The complete cycle of view, move and view again must be completed in $^1\!/_{24}$ second, calling for a type of mechanism which in 1908 was definitely a matter of high technology.

The intermittent movement of film must be confined to the small part of film in the area of the viewing gate. Any attempt to make a whole reel of film move in this way would certainly cause the film to break, and it is only by confining the movement to a small and therefore light piece of film that it can be made to work at all. A more pressing reason for confining the intermittent movement came when sound tracks were first used, requiring the film to be passed at a perfectly steady rate through the sound gate in order to avoid the rough buzz heard from the loudspeaker if the film is moving jerkily at this stage.

An important feature of movie projection is therefore the formation of loops, (Figure 3.1). By looping the film before and after its passage through the gate enough slack is provided to allow

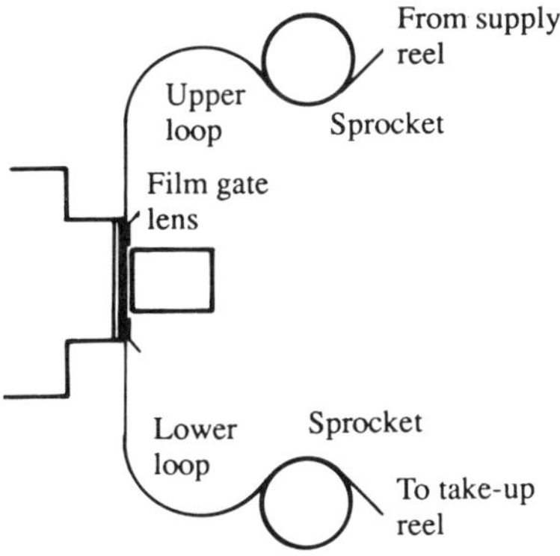

Figure 3.1   *The two loops of film which must be present on a 16 mm projector to allow intermittent movement of film through the gate*

for intermittent movement, while permitting film to be fed at a steady rate before and after the loops. One of the main causes of problems with movie film is 'losing a loop', often due to damaged perforations (see later). The Edison film camera and projector used 35 mm film perforated at each edge with four perforations per frame and a picture size of 1″ × 0.75″. This 35 mm film size has persisted and is still the basis of many commercial films as well as being the film for the ubiquitous 35 mm stills camera; the frame size for stills is increased to 36 × 24 mm by using the film sideways. The other remarkable survivor from the past is the first successful type of mechanism for intermittent film movement; the Maltese cross mechanism, illustrated in Figure 3.2. A pin held on a revolving wheel strikes the Maltese cross wheel, turning it through a quarter of a revolution, and leaving it in place until the pin engages in the next slot of the Maltese cross. The film itself has perforations (sprocket-holes) of rectangular section at each edge so it can be positively located on a sprocketed wheel on the same shaft as the Maltese cross. The size of the pin-wheel is chosen so the mechanism spends one quarter of the time of a cycle in moving the film while three quarters of the cycle time is spent viewing the film. For the standard 24 frames per second rate of projection, this implies film is being moved on in the time of $1/96$ second.

The same intermittent movement is used to turn the rotary shutter (Figure 3.3) whose position relative to the Maltese cross is arranged so the solid portion of the shutter is over the light beam the Maltese cross rotates. This ensures the eye sees only the stationary image. The shutter is arranged with two blades, a primary blade which interrupts the light as the film is advanced, and a secondary blade which interrupts it half-way between the film advance intervals. By using the extra blade, flicker is further reduced as light is interrupted at a rate of 48 times per second while film is projected at 24 frames per second. Some projectors use three-bladed shutters to reduce flicker still further.

The original type of Maltese cross mechanism is still used in 35 mm and 70 mm projectors, and in some telecine projectors used for converting film images into TV signals. The clatter of the pin into the slot makes the sound which many enthusiasts associate with the cinema, but this sound becomes a problem if the projector is in the same room as the screen. The essence of commercial film projection is that the projector is in a sound-proof room, with the beam of light passing though a window to the screen. This is seldom possible for smaller-scale film projection use as a visual aid.

The film gauge for professional cinema applications was for many years 35 mm, producing a frame of 36 mm × 24 mm, though the larger 70 mm format has been extensively used, particularly for wide-screen work. A complication of professional 35 mm film has been the use of anamorphic lenses on exposure, which distort the picture on the film in such a way as to provide wide-screen presentation when the film is projected with another lens of the same type. Such films do not satisfactorily adapt to 16 mm and cause problems when adapted for TV, when the choice is either to show the film as a narrow strip across the width of the screen or to

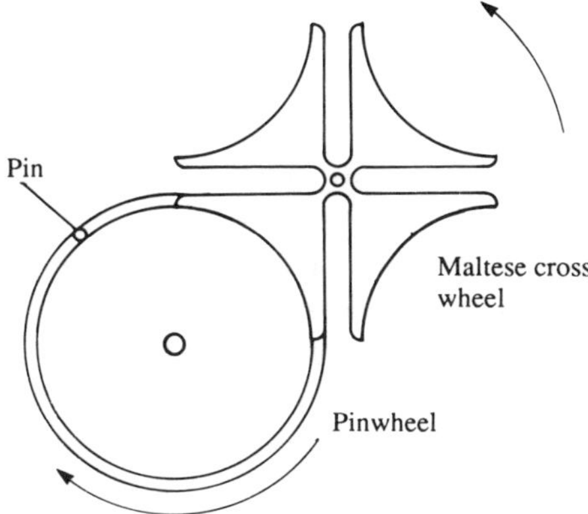

Figure 3.2   *The old-style Maltese-cross mechanism which has been used on large 35 mm projectors since the start of movie pictures*

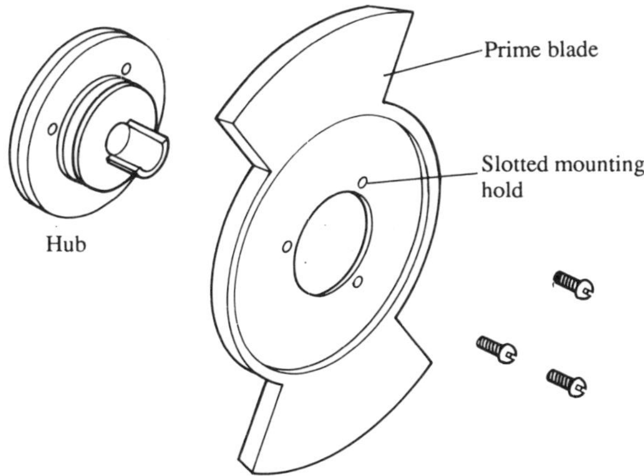

Figure 3.3   *Shutter-blades of a typical 16 mm projector. Three-bladed shutters are also used. The primary blade is the one that interrupts the light while the film is being moved*

show full-height but selecting only a part of the width in any given scene.

The full 35 mm size has always been too large for AV use, and the smaller 16 mm gauge has been standardized for AV applications and also for some TV uses and for commercial filming. At one time 9.5 mm was a possible contender and the Super 8 mm format was tried in later years, but the vast bulk of material of interest to educational and industrial users is available on 16 mm film. No other gauge has ever become a serious challenger to 16 mm for educational uses. Smaller gauge film was, in fact, the main impetus to the use of film as a visual aid. Then known as 'sub-standard gauge', 16 mm was available very early in the life of cine-photography, and older 16 mm projectors use the same type of Maltese cross mechanism as larger 35 mm machines. The film, however, is not nowadays comparable. When 16 mm film first became available it was a miniature replica of 35 mm film, with a row of sprocket holes along each edge. This was silent film, and when sound film became a reality the method used for 35 mm film could not be so easily applied to 16 mm. The sound-track (see later) for 35 mm film uses a narrow band of film between the sprocket holes and picture frames of the picture. On 16 mm this strip is too small, so the solution adopted was to make the film with only one edge of sprocket-holes using the other, plain, edge for the sound track.

This method works well, but it requires the intermittent movement to work along one edge of the film only. The Maltese cross mechanism with sprocketed wheels is not so suitable and, in addition, it is not a mechanism which can be easily adjusted to compensate for wear. This has led to a very different type of intermittent mechanism, used for 16 mm projectors since the 1950s.

This mechanism uses a claw operated by a cam (Figure 3.4) with the claw arranged to engage in more than one of the film's

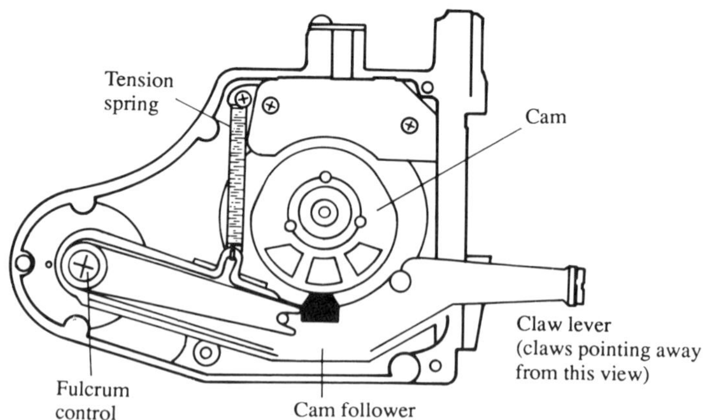

Figure 3.4   *Cam and claw movement of a 16 mm projector. Thickness of the cam is used to operate the protusion and retraction of the claw, and profile of the cam moves the claw up and down*

sprocket-holes. The claw is retracted while the shutter is open, projecting a stationary image, then pushed forward to engage the sprockets and moved down to pull the film down by one frame before being retracted again. During this last stage of movement a shutter blanks off light from the film gate. This type of mechanism is quieter than the Maltese cross mechanism and easier to adjust. Use of a double-claw (Figure 3.5) greatly reduces incidences of lost-loops caused by tearing of a sprocket-hole, particularly on the one-sided sprocket holes of the 16 mm film. Claw movement can be obtained from one cam, using the edge of the cam to operate up and down movement of the claw, and variations in its thickness to operate claw retraction and protrusion.

The critical part of the projector is the gate (Figure 3.6) in which

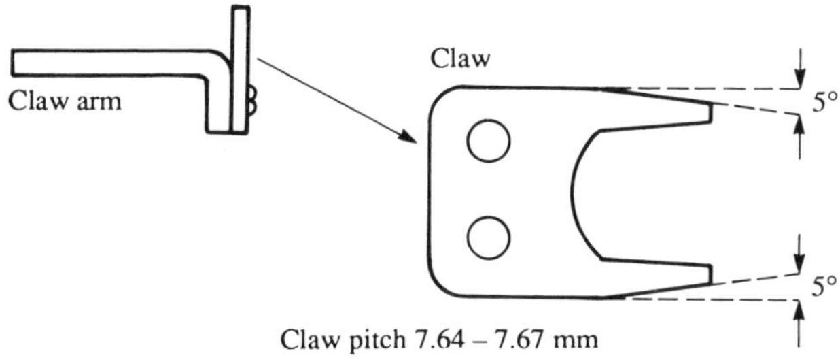

Figure 3.5 *A typical double-claw, showing the slight angle of the outer edges essential to ensure smooth entry into the film sprockets*

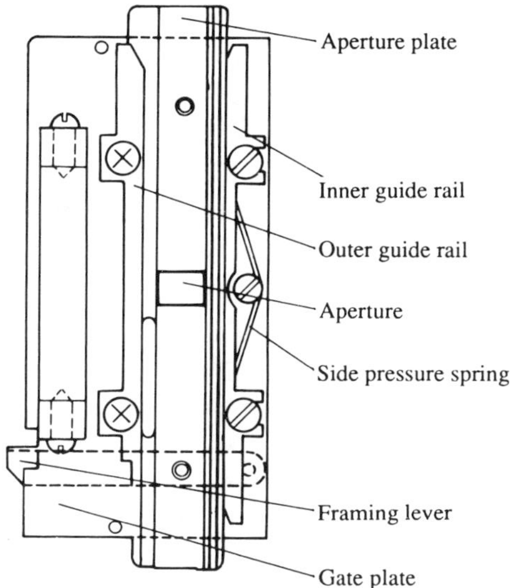

Figure 3.6 *A projector gate. Fit of the film in this gate is important; it must be neither too slack nor too tight. The claw emerges from the slot just under the aperture and (in this view) to its left*

the film slides, registering each frame in turn with the aperture from which the light beam emerges. Dimensions of the side channels of the gate are held to very tight limits. If the gate is tight, film sticks and the claw rips the sprocket holes. This may be followed by a fire as the light beam heats up the film excessively. If the film is too slack, the one-sided action of the claw can twist the film slightly, making it jam or causing it to move slightly in the projection time. Incidentally, still projection of individual frames can be achieved on some projectors by using a clutch mechanism which stops the claw action and, when this is done, a heat-absorbing glass filter is placed between the gate and the lamp-house. Even with this precaution, however, film deteriorates rapidly if the'still picture is held for too long.

Standard dimensions for 16 mm sound film are shown in Figure 3.7. Frame size is 10.22×7.37 mm, which amounts to an area of 75.32 mm^2 as compared to the 484 mm^2 of 35 mm cine film. This is a factor of about 6.5, implying (with the same screen magnification) 16 mm film is 6.5 times less bright. Even if we allow that screen size of 16 mm film is considerably smaller, this is a considerable difference. Film is, quite literally, 35 mm film sliced down the middle so there is only one set of sprocket holes along one edge. The other edge is reserved for the sound track, which can be optical or magnetic. When a magnetic sound track is used, this edge is 'striped' with magnetic oxide and the recording is made using the normal methods of tape recording. Optical tracks, on the other hand, are of the variable-area type in which the width of the track is proportional to the amplitude of the sound.

The use of an optical soundtrack on one side of the film makes alignment of film in the sound pickup portion of the projector critical, because if the film frames impinge on the sound pickup the result is a loud hum or rasping noise. Principles of the optical sound pickup system are illustrated in Figure 3.8, which also shows the form of the variable-area soundtrack. A light beam obtained either from the main lamphouse or, more commonly, from a subsidiary bulb (the exciter lamp)is focused through a small aperture on to the sound-track area of the film. The beam passing through is picked up by a photoelectric cell, whose electrical signal output.is proportional to the amount of light reaching it. Thus, the photoelectrical cell converts the variations in width of the sound-track area, detected as variations in the brightness of the light-beam, into electrical voltage variations. Early soundtrack systems used a track of constant width in which the sound was modulated by variations in the darkness (density) of the film. This

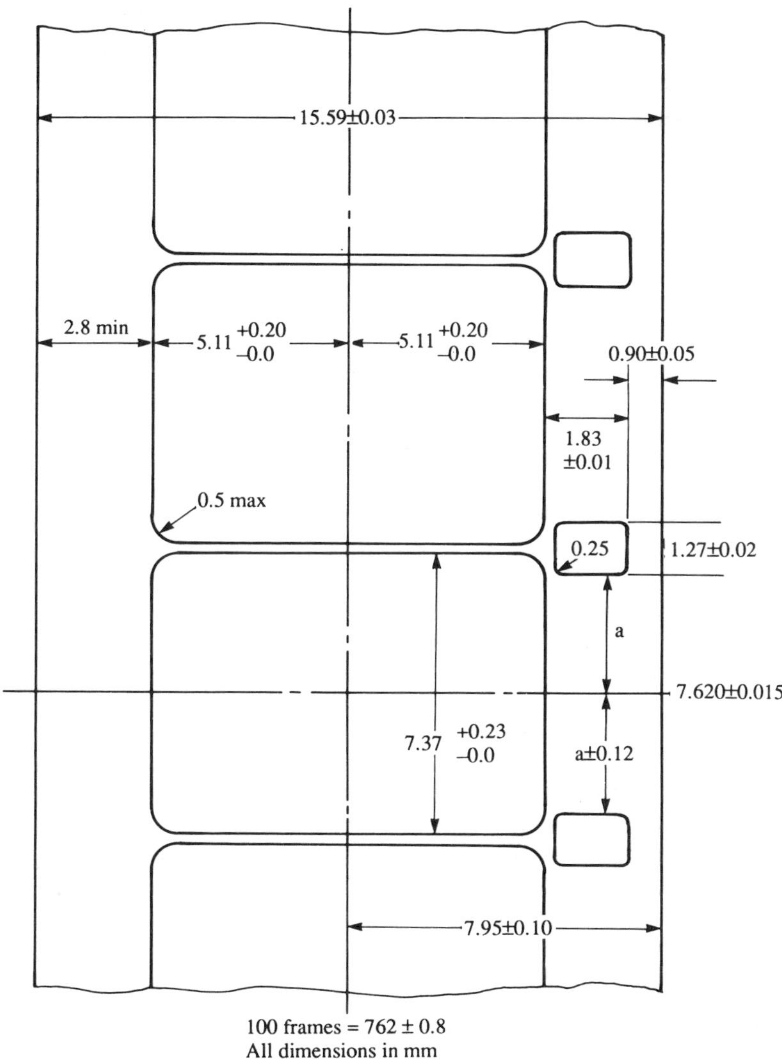

Figure 3.7 *Dimensions of 16 mm film – unlike 35 mm film, this uses one sprocket-hole per frame*

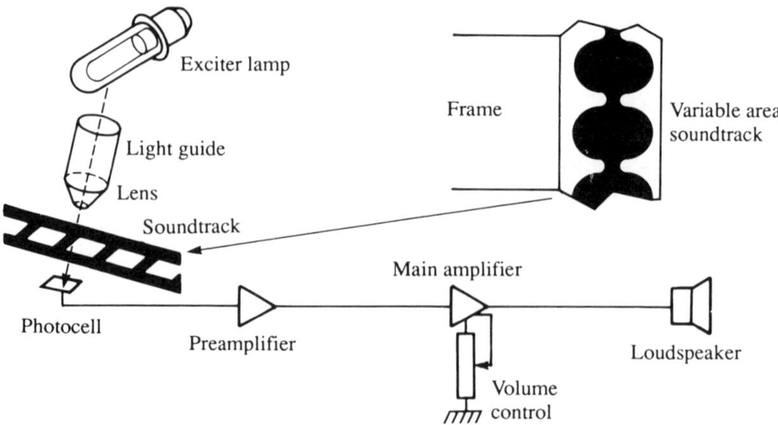

Figure 3.8 *Principles of optical sound. The amount of light reaching the photocell is constantly varying because of the shape of the soundtrack, and this variation is converted into electrical signals, amplified to operate the loudspeaker*

system has been abandoned because of its sensitivity to the processing of the film (density can be altered in processing, shape can not) and to scratching, and the variable-width type of system has been totally predominant since 1950.

The output from the photoelectric cell is taken to the amplifier system, whose output is the loudspeaker socket of the projector. Most units are devised with a loudspeaker built into the carrying case and a generous supply of connecting cable already attached and ready to plug into the loudspeaker socket. Amplifier controls on older machines are usually simple volume and tone controls with no provision for bass cut and treble boost, which could improve the intelligibility of speech on many films. Later machines, such as the EIKI *Model N* feature separate treble and bass controls, with an amplifier system which caters for either optical or magnetic sound.

The position of the sound head is important. The film reaching the sound head has already passed through the gate, and the intermittent movement of the loop has been smoothed out by driving the film through a set of sprocketed wheels which incorporate a flywheel so as to keep the rotational speed constant. This means sound for any particular frame is placed on the film ahead of the frame itself, – standardized distance is 26 frames, corresponding to a time of just over one second of film movement.

From this it follows that if a frame is removed, sound and picture will not match up until 26 frames later. Removal of one single frame throws the synchronization out by only $^1/_{24}$ second, an amount which is not serious, but on some old films which have been extensively repaired mismatch of sound and picture can be very noticeable. Equally bad mismatch can be obtained if lower loop size is incorrect, because the lower loop contains part of the 26 frame length which has to be allowed between gate and sound head. A mis-match of a few frames is not very important, but if the loop is excessively large poor synchronization is particularly obvious as on-screen characters start to speak, the lip-sync problem. If, on the other hand, the lower loop is lost or drastically shortened some of the intermittent movement through the gate affects movement past the sound head, causing very distorted sound and the film is likely to break at the first weak point (often a previous break point).

This means loop formation is critical, and one of the most useful developments in 16 mm projection is the auto-threading projector which, given undamaged film to work with, leads the film through its correct path and makes loops of the correct size (the precise size is adjustable). Self-threading obviates most of the problems which plagued use of movie projectors in the past, but it relies heavily on film being undamaged because film is pulled into place using the sprocket holes on the first few dozen frames. If these sprocket holes are torn, automatic loading is almost impossible to carry out, and there is often little or no provision for threading film by hand.

User guide

The first essential action in using 16 mm movie film is to check film condition, particularly the first critical set of (blank) frames which are used in automatic threading. If these frames are damaged, they should be removed and/or replaced – damage in this sense means tearing of the sprocket holes in particular. AV departments should keep a supply of good blank 16 mm film which can be spliced in to replace damaged *leader*, as this section of the film is called.

If the AV department has a 16 mm editor, film can be viewed on a small screen to see what is available and, more important, checked for weak splices or breaks. Films which are extensively borrowed may be posted out with little or no inspection, and if the previous user has been careless the film may prove to be in poor condition so the AV technician must make repairs before the film can be used. Some film libraries are scrupulous about film quality and never allow a poor copy to go out, others are either less careful or have less time and after a few years of borrowing films you are likely to know which films are most likely to cause problems.

Before even contemplating a session of film projection, you should take some time to know the projector. In particular, you need to know what settings have to be made and how to operate the auto-threading system, because a half-darkened, noisy class-room is no place to have to learn these points. This stage is particularly important if the AV department offers several different models of 16 mm projectors of very different ages.

Most projectors allow for the use of 18 or 24 frames per second (fps) projection rates. The 18 fps rate is retained only for the very few silent films still existing in their original form – most silent films transferred to 16 mm have also been converted to be shown at 24 fps, but if you encounter a film which is silent and on which movement is unnaturally fast and jerky this indicates the 18 fps rate is intended. Most films, and certainly all sound films, use the 24 fps rate setting.

More modern projectors also feature a switch for setting either magnetic or optical sound track. Older films all use optical sound tracks, and this will be the normal setting for the sound switch. Check your film well in advance to see if a magnetic recording is used – this ought to be notified with the film and if a film has been provided with magnetic sound track a corresponding projector is required.

Most projectors allow use of a microphone, so the user can superimpose a commentary over that on the film, or add a commentary to a silent film. On some machines a separate volume control is included for this purpose, allowing commentary and film sound to be mixed; on other machines the selection switch for optical or magnetic sound has a third position for microphone. This latter system does not allow mixing of sounds and you need to be careful, when returning to film sound, you select the correct switch position, optical or magnetic according to the method of recording used on the film.

Another facility which magnetic sound-track film presents is the possibility of recording a sound-track. This would not be used on film with a sound-track of its own, but for some silent films addition of a custom-made sound track is useful. Silent film can be 'striped', using a self-adhesive magnetic tape, so a soundtrack can be recorded while the film is playing. Very great care should be taken to ensure an existing magnetic soundtrack on a hired film is not wiped by recording in error – another good reason for being familiar with the projector before using it.

The other control you need to be aware of is the framing control, usually positioned at the gate. This allows small adjustments to be made in the positioning of the film-advance claw and will have to be used if, on projection, you can see part of another frame. Turning the framing control wheel allows the normal framing to be restored. Abnormal framing may simply be because someone has twiddled the control before the film started, but it can also indicate abnormal stretching of the film and should be reported to the technician.

The other side of the threading problem is that if your own projector tears a film you should either have it repaired or send a note back with the film explaining what has happened. Film libraries are not likely to hold such accidents against you, and would rather know when a film has been damaged than have the comment made by the next user. When the film is borrowed from your own AV department, it ought to be checked before being re-issued.

Other aspects of movie projection follow closely the use of slide projectors. Darkening of the room is the first priority, and is more critical for movie work than for slide projection. Back-projection using mirrors is sometimes used as an expedient for rooms which cannot be satisfactorily darkened, but a good back-projection set-up really requires a rigid structure of screen and mirrors to be made, and is not likely to have a long life if it has to be moved around (or is left unguarded). Direct back-projection cannot be used because, unlike slides, movie film cannot be put into the gate the wrong way round or upside down – direct back projection would produce laterally reversed (left-for-right) images.

Positioning of a movie projector is more critical than that of a still type and makes the use of appropriate lenses even more important. The movie projector is larger, noisier and more obtrusive than the slide projector, and space has to be left for the two film reels to revolve unobstructed. Though the cooling fan can be made reasonably quiet the characteristic clatter of the intermittent drive mechanism is always present, even in these days of quiet plastics gears. The preferred arrangement is to have the projector as far back in the room and distant from the bulk of the viewers as possible. The loudspeaker should be placed high and close to the screen (but never behind the screen). Any obstacles in the path from loudspeaker to ear will make the received sound less intelligible (assuming it was intelligible to start with) by reducing the treble content. The screen also needs to be placed high enough to be clearly seen by all viewers, and tilted if necessary to minimize keystone effect. If the cable between the projector and the loudspeaker is not long enough an extension should be made.

One very useful accessory for movie projection is a remote control for room lights. Very few rooms are likely to permit this for the main lights, but simply having a lamp close to the projector which can be controlled by the operator is enough to allow everyone to see where main light switches are. As movie projection calls for a higher standard of blackout in the room than other forms of visual aids, it is important not to be left blundering about in darkness if the film loop should fail or, worse, in the event of an emergency such as a fire in the building. Light control of this kind also allows the operator to get the film started then switch out the last lights rather than depend on someone else switching off the lights, or to risk trying to walk from the light switches to the projector in suddenly imposed darkness.

During projection, it may be necessary to alter volume control of the sound system. One perennial problem is that the projector operator usually sets sound volume level so it can be heard over the noise of the projector. For most of the audience, however, sitting further away from the projector, this volume is uncomfortably loud. This is, perhaps, not so noticeable in schools in which a significant proportion of pupils are partially deaf from the combined effect of discos and portable cassette players, but for adult audiences excessive volume is a decided irritant. Combined with poor quality of sound often obtained from movie film (or video), excessive volume can distract attention from the message of the film. Audio quality of magnetic striped film, incidentally, is considerably better than an optical soundtrack's, but even a magnetic soundtrack is fairly poor compared with sound from a cassette made using Dolby noise-reduction. Though Dolby methods are by now extensively used in the commercial cinema, it is most unlikely now to be applied to the smaller gauges.

The two problems most likely to arise during projection are bulb failure and loop loss. Bulb failure inevitably causes a serious delay, because movie projectors do not generally allow rapid bulb change, and it is usually necessary to find the AV technician and wait until the bulb change can be carried out. If a loop is lost, adjustment of threading can often be difficult, because the auto-threading mechanism makes it difficult to adjust threading manually. The film has to be taken off the sprocketed wheels, adjusted so two loops of the correct size are formed, then placed on to the sprocketed wheels again. The projector should not be started until the film has been wound on by hand for a few frames to ensure loops are maintained.

When a film loses a loop in the middle of projection, the technician should be informed so the film can be repaired. Loops are lost mainly because of faulty perforations or bad splicing, and if the film is simply rewound and returned, it will quite certainly fail in the same way the next time it is projected. It helps considerably if you can provide some indication of where the film failed, either in terms of running time (ten minutes after starting) or by some prominent visual mark on the film (just after launch of the liner) so the technician does not have to examine the whole of the film in detail to find the offending part. If loss of a loop cannot be attributed to faults in the film, the claw mechanism must be checked and, if necessary, overhauled.

After projection, film should immediately be wound back. This allows the projector to cool down, because on rewind the lamp is not in use but the fan is, and this forced cooling is beneficial. The only exception to this rule occurs when a film is supplied on more than one reel, when you need to make a rapid change over from one reel to the other, and rewinding imposes too long an interval. In such a case, remove the full take-up spool, move the empty spool to the take-up position, and load in the second (or subsequent reel, and resume projection, leaving both reels to be rewound later. During rewind, look for breaks in the film.

Technical maintenance

Unlike a slide projector, the movie projector requires maintenance each time it has been used. The movement of the film through the gate inevitably shaves off pieces of emulsion or other film coating, and this debris accumulates all over the projector but particularly in the gate. The gate must therefore be dismantled and thoroughly cleaned each time the projector is used, and the whole film path should be checked as well. Even a tiny projection of hard emulsion in the path of an auto-threading projector can cause trouble if it catches an edge of film or is caught in a sprocket hole while the film is being laced through. A stiff brush is by far the best method of cleaning the gate, and hard deposits should be scraped away with a plastic spatula, with the last remains removed by polishing with a cloth. Never risk scratching the gate with any metal object, it is as vulnerable as the head of an audio or video recorder.

Bulb failure in movie projectors occurs at a higher rate than in slide projectors because of the greater vibration experienced in the movie projector. This makes it more important to try to adopt a system of periodic replacement of bulbs on the basis of time in service, because the cost of replacing a bulb that is still working is less than the cost involved in making a replacement in the middle of a film-show. Movie projectors seldom make any provision for running the bulb at a reduced rate; the problem in all movie projectors is to get sufficient illumination on to the film. A modification placing a resistance in series with the bulb at the time of switch-on is shown in Figure 3.9 which, if incorporated, will greatly increase life of the bulb. The objection to this is that each user of the projector then has to be instructed in the use of the limiting resistor and its by-pass switch. The exciter bulb for the soundtrack seldom fails, but it should not be ignored as a result.

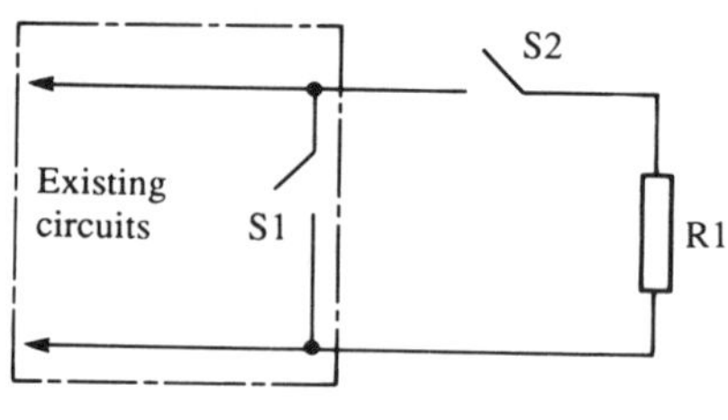

S1 Existing lamp switch
S2 New on/off switch for lamp
R1 0.1 ohm 15W resistor

The lamp is initially switched on with S2, and S1 is
then used to increase the brightness to its full extent

Figure 3.9   *Using a two-position switch for bulb brightness. If the bulb is always switched on in the LOW position its life is greatly extended. The resistor can be bought or made from resistance wire on a ceramic former*

As 16 mm projectors are not so common, more details about their maintenance are appropriate here, particularly as maintenance manuals are not easy to come by. A few specialist repairers possess manuals but are not likely to loan them out and some swapping and photocopying with other AV centres can be advantageous. Importers of projectors can usually find manuals, but this can take time because of the huge number of different models and variations in each model for country of use.

The trouble-finding charts of Tables 3.1 to 3.4 are for the popular EIKI N series, but are suited to a large range of modern 16 mm projectors. Mechanical problems, other than those caused by defects in the film itself, are normally dealt with by adjustment and attention to lubrication. The most critical adjustments are to the claw mechanism, in particular to claw protrusion which decreases as the mechanism wears. Adjustment of claw protrusion is best done using a jig which can be obtained from the manufacturers or made up if suitable workshop facilities are available. Adjustment for claw pitch should also be checked at intervals, because incorrect adjustment (due mainly to wear) causes the claw to damage sprocket holes.

Table 3.1 Fault-finding table for 16 mm projectors – electrical faults

Problem	*Cause(s)*
Motor will not run in forward direction	Motor defective. Motor connections open circuit. Faulty capacitor. Forward microswitch or operating cam faulty. Motor thermal switch open
Motor runs but lamp does not illuminate	Lamp blown out. Lamp not connected. Micro-switch faulty. Transformer connection open circuit or faulty transformer
Motor runs forward but will not reverse	Microswitch defective or motor defective
Function switch does not follow correct sequence	Loose switch knob. Loose switch cam. Incorrect installation of reverse or take-up clutch cam.

Electrical faults: This table concentrates on internal faults, assuming such items as fuses and cables have been checked. Presence of the pilot light indicates power is present.

Table 3.2 Fault-finding table for 16 mm projectors – automatic threading faults

Problem	*Cause(s)*
Self-thread lever does not set mechanism	Cam bracket loose on guide shaft. Main interlock bracket binding or not latching with release hook
Leader being chewed by sprockets	Film wrong way round. Insufficient clearance between sprocket-plate and film shoe. Film path and sprocket drive misaligned. Film leader is soft
Leader jams in first guide	Leader not correctly trimmed. End of film is curled or twisted. Soft leader. Shaft of first guide loose or bent. Aperture plate misaligned
Tip of leader enters under inner guide rail	Inner guide rail bent or misaligned. Loose screws on guide rail. Leader severely twisted or curled
Film jams in gate	Leader incorrectly trimmed, curled or twisted. Excessive clearance on first film shoe. Dirty or obstructed gate. Inner guide rail has insufficient travel. Side pressure spring tension excessive. Film shoe not retracting during threading
Chattering noise	Claw not completely retracted
Clicking noise	Claw hitting shutter-blade
Film goes over loop setter roller	Loop-setter too low. Film badly curled
Film goes under third guide, or comes out	Second or third film guide defective or badly aligned. Film badly curled
Film will thread over sound drum	Rough spot on lamp-house casing. Pinch roller not completely raised, stuck or misaligned
Film stops at second sprocket	Rough spot on lamp-house casing. Sprocket teeth or cover plate loose. Tension guide and roller misaligned. Sprocket shoe too tight
Film comes out of second sprocket	Sprocket plate too loose. Sprocket shoe spring weak or broken
Self-thread does not release correctly	Broken or weak release bracket spring. Pin of release bracket is binding

Table 3.3 Fault-finding table for 16 mm projectors – mechanical faults

Problem	Cause(s)
Motor runs, film does not advance	Still picture selected. Slack or broken motor belt Loose washer on cam tank plate. Motor pulley loose Main drive belt broken.
Film speed incorrect	Belt incorrectly fitted on pulleys. Wrong pulleys fitted to machine
Film comes out near sound drum	Pinch roller binding. Roller and tension guide out of alignment. Second sprocket shoe not seating
Excessive tension at forward wind pulley	Cork clutch dry. Arm belt or pulley dirty or sticky
Excessive tension at reverse wind pulley	Reverse clutch spring too strong
Forward take-up poor or absent	Arm belt broken, loose or oily. Defective ball bearing. Loose spindle shaft. Take-up pulley binding. Over-greased take-up pulley. Clutch cam defective. Film too loose on take-up reel
Reverse take-up poor or absent	Arm belt broken, loose or oily. Reverse belt oily or broken. Clutch cam not engaging. Reverse clutch spring weak
Insufficient back tension of supply reel	Reverse clutch cover pulley binding or clutch cam not releasing. No end play for clutch pulley. Clutch cam seating defective. Clutch cam binding between clutch cover pulley and spacer
Insufficient back tension of take-up reel in reverse winding	Drive gear binding or no end play on drive pulley shaft. Clutch cam defective or badly seated. Drive gear or clutch cover defective
Loop setting roller continues to operate during forward run	Damaged film. Weak gear spring. Incorrect position of loop setter roller. Loose second sprocket plate. Lower loop too small. Insufficient claw protrusion or claw pitch. Broken claw. Incorrect clearance between loop setting gear and main drive belt. Loop setting gear or drive belt broken or defective. Low tension in film shoe springs

Problem	*Cause(s)*
Film flaps against loop setting roller during forward movement	Binding loop-setter gear shaft or loop-setter gear spring tension excessive
Lower loop lost when reverse engaged	Reverse rubber roller not driving the flywheel set collar. Reverse rubber roller binding. Second sprocket not operating
Under loop lost when reverse engaged	Damaged film. Excessive clearance on shoe of first sprocket. Incorrect shoe protrusion
Noisy projection action even with undamaged film	Upper loop too small. Film striking loop-setting roller. Dirty gate. Loose claw. Incorrect claw protrusion. Inner guide rail binding. Film show damaged. Claw position incorrect. Cam follower spring weak or broken
Ghosting on picture	Incorrect shutter blade postion
Noisy on reverse only	Claw position incorrect, worn cam follower or pin
Noise on still picture	Motor pulley misaligned. Shutter pulley binding on shaft
Film will not stop when still selected	Shutter pulley seized. Shoulder screws on still-picture lever loose
Film burns when still picture selected	Still picture lever not completely depressed. Heat filter broken or misaligned
No rewind, or poor rewind action	Cork on take-up spindle clutch too dry or tight. Supply arm belt broken, stretched or oily. Rewind gears faulty. Rewind arm tension spring broken. Take-up pulley binding on shaft
Noisy rewind	Rewind gears not fully engaged, worn or defective
Uneven focus	Dirty gate. Film shoe binding or not seated. Inner guide rail binding. Lens holder misaligned.

Table 3.4 Fault-finding table for 16mm projectors – lamp faults

Problem	*Cause(s)*
Lamp life abnormally short	Socket in bad condition. Cooling restricted. Lamp of incorrect type. High AC line voltage
Uneven or insufficient screen illumination	Lamp badly seated. Obstruction in light path. Lamp of incorrect type. Function switch in LOW position. Lens dirty or defective. Low AC line voltage. Shutter not operating correctly

The other factor making movie projection so much more work than still projection is the maintenance of films. Universal adoption of self-threading projectors makes it vitally important to ensure sprocket holes on the film are in perfect condition. In the days of manually laced projectors, condition of the sprocket holes at the start of the film was almost irrelevant, because the user would wind that initial piece of film around the take-up reel and then lace the remainder through the gate and the sprocketed wheels. With automatic threading, the first part of the film must be perfect, and if damage has occurred, a new leader section should be spliced in.

Splicing can be carried out using film cement or sprocketed tapes. Use of tapes is much faster, splices are strong and, more important, unobtrusive. Splicing tapes can be obtained which do not extend as far as the sound-track so that a magnetic sound track is not obstructed by the tape – the optical type is not affected because the tape is transparent. In use the broken ends of the film are held in a splicer which locates precisely on to the sprocket-holes, and each broken section is cut along the frame-line (the space between adjacent frames). As film seldom obligingly breaks along this line, cutting is usually needed on both halves and involves losing a frame. With the film cut and the two ends located in the splicer, tape is then placed on the sprocket holes each side of the cut and smoothed over the film.

Though some modern film materials *must* be repaired with tape, many users prefer the older method of using cellulose acetate cement. The film is prepared by cutting as before, but for cementing repairs the splicer should incorporate a scraper, preferably with a serrated edge. The scraper is used to make each edge wedge-shaped in side view, so the two pieces of film can be cemented without greatly increasing the film thickness at the splice. If this type of splice cannot be carried out the usual method is to cut the frames on each side of the frame line, so that the join can be made in the small area between frames (a mask-line splice). This type of splice is thicker than the scraped type, and so clicks as it goes through the gate of the projector. It is also a weak point on the film, though a well-made splice, allowed to harden properly, is remarkably strong.

In general, tape splicing is easier and the only drawback is cost of tape if a large amount of splicing is undertaken. If good care is taken of films, splicing should not be necessary often, particularly now 16 mm film is not the main medium for showing moving pictures, and tape splicing justifies the cost of tape patches in the time saving possible.

Hints and tips

- Old 16 mm film can be transferred to video. This should be done professionally, using a flying-spot scanner rather than by makeshift methods based on back-projection to a screen which faces a video camera. This simple type of transfer is effective for slides, but with movies causes problems because frame rates of 24 per second for film and 25 per second for video cannot be synchronized. Purchase of a flying-spot scanner is most unlikely to be justifiable unless you have a complete library of films to transfer.
- Irreplacable old films can be coated with a protective and lubricating varnish, which considerably reduces breakage problems.
- Use only silicone oils or grease on plastic parts.
- Avoid using trichloroethylene as a cleaner – it will cause most plastics to swell.
- Most metal bearings can be lubricated with mineral oil or greased with molybdenum disulphide grease.
- If there is persistent hum on playback of magnetic tape, check for misplacement of the humbucker coil, a small coil of wire near the pickup head which is used to counteract the pickup of hum by the head.
- Most projectors are far from silent, but excessive noise often indicates wear. Look in particular for fan blades striking the claw arm or the claw hitting the gate.
- If the film shows faint ghost images, this is almost certainly an indication that shutter blades are incorrectly aligned. The blade

assembly is located on the hub by screws, but slots in the blade arms allow the blades to be rotated slightly either way. This setting has to be found by trial and error.

- Prolonged use of still-frame action causes heat-damage to the film and quickly wears out bushes in the projector.

4

Tape and cassette recording equipment

Audio AV aids equipment is nowadays centred around various types of cassette recorders, of which language laboratories account for some 90% of equipment use. The convenience of cassette recorders ensured that they replaced the old-style school gramophone virtually as soon as reliable cassette mechanisms were made, but replacement of open-reel units by cassettes in language laboratories is slower. Open-reel recorders are still in use as master-replay units in language laboratories but convenience of cassette equipment together with developments in cassette reproduction technology are steadily eroding the place of open-reel tape. In addition, the number of master tapes that can be bought ready-made on open reel tape is also diminishing in favour of the cassette type.

Reasons for the dominance of the audio cassette are not difficult to see – a recording medium which is compact, can be used over and over again or protected against accidental recording, yet give a reasonable standard of reproduction represents a considerable advance for educational use over any of the 'black vinyl' methods of recording of the past. Open-reel tape systems offered the same advantages, but with drawbacks of requiring the user to lace tape into the recorder, and the inevitable spillage of tape if a reel were dropped or if a reel jammed on the machine. The compact nature of cassette recorders themselves is a mixed blessing, because the more compact a piece of equipment is the more difficult it becomes to keep it in its place, or to keep it at all.

Another application of cassette audio equipment in educational work is to time-shift radio broadcasts. The problem of copyright is dealt with by allowing such recordings on the condition tapes are wiped within a reasonable time of making the recording. Increasing use of television has to some extent diminished the importance of sound recording of educational broadcasts, which is unfortunate, because radio broadcasts have always been rich feasts for the imagination, in contrast to the more plodding approach of television. The TV image appears to replace the imagination rather than stimulate it.

Another approach never possible in the days of the gramophone record is to use the cassette recorder as a reporting tool, recording birdsong, industrial noises or verbal interviews with local worthies. This type of use, from primary school level upwards, is a stimulant to curiosity, which also serves to make a valuable link between the classroom and the outside world. Problems which attended use of microphones with the first generation of cassette recorders are now long forgotten, and an excellent standard of recording and reproduction is now achieved with little need for technical expertise.

The use of tape recording equipment overcomes the old problem of learning a language, where the student never hears a language spoken by anyone other than the teacher or another student, and certainly never heard his/her own voice reproduced. Use of (particularly) cassettes makes it possible for language students to hear a wide range of accents of native speakers of the chosen language, something that was very difficult to achieve in earlier generations. They can also hear their own attempts at pronunciation, something which is of quite inestimable value. Normally, we hear our own voices partly through our ears and partly by the conduction of sound through the bones in the head. This gives a totally different impression of your voice to another listener. Interactive use of cassette recorders in the language laboratory allows each student to be immersed in the spoken language without the embarrassment of speaking out loud in a class, and also proceeding at his/her own pace. The main danger is that formal grammar can be lost in such a situation, a point unanswered by the view that as English grammar is no longer taught there is little point in teaching the syntax of any other language. A combination of conversation, emphasis of phrases and grammatical points is an essential feature of good language laboratory use and encouraged by good design of the language laboratory.

Principles

Tape recording, whether on open-reel tape or on the small enclosed reels of the cassette, is the use of a magnetic material for recording. Just as the gramophone record is a mechanical system, using soundwaves to control the size and shape of a groove in a soft material, tape recording is a magnetic system, which magnetizes the tape with a pattern representing size and shape of the soundwave.

The idea of magnetic recording is almost as old as that of mechanical recording, and a practical recorder using steel wire was first exhibited by Valdemar Poulsen in 1898. Basic principles of the modern tape, shown in Figure 4.1, are simple enough. The sound waves are converted by microphone into electrical waves and the varying electric current of a wave causes the elctromagnet to set up a varying magnetic field. This magnetizes the magnetically-coated tape, and, as the tape is moving, each part of the tape is left magnetized to an extent reflecting the magnetism caused by some part of a sound wave. When the process is reversed, passing the tape with its variations in magnetism past another electromagnet, a corresponding signal is obtained from the electromagnet directly analogous to those of the original wave, which can be amplified and used to drive a loudspeaker.

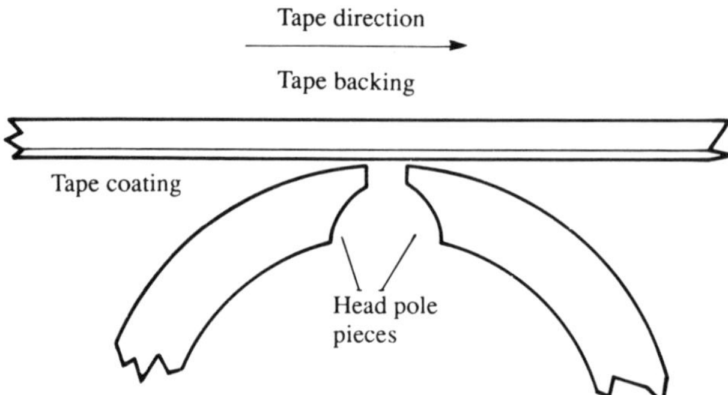

Figure 4.1 *Principles of the recording/replay head of a tape recorder. The original Poulsen device used steel wire rather than tape, but was essentially identical*

If it were only as simple as all that, tape might have replaced mechanical recording many decades ago. As it was, magnetic recording was used to some extent for dictaphones in the 1920s, but its use had all but died out by the time World War II broke out. Tape systems we use today were developed in war-time Germany, mainly by BASF, and the technology astonished the first Allied engineers who examined the equipment. As a result, patents were declared war booty and designs were adopted by manufacturers all over the world after 1945.

The main problem of recording on tape is that the simple system, as outlined here does not work satisfactorily. The relationship between the retained magnetism on tape and signal currents in the electromagnet is a very complicated one, which makes replayed sound from such a system very distorted. Ideally, the relationship as shown by a graph should be a straight line, in practice it is a curve (Figure 4.2) which shows the tape retains virtually nothing when signal levels are low.

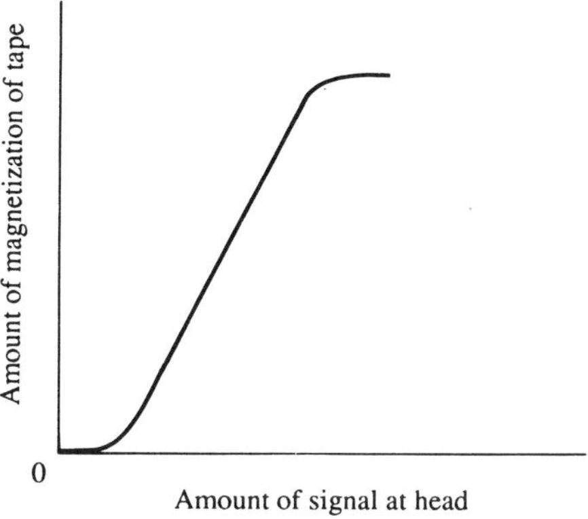

Figure 4.2   *Amount of magnetization of tape is not perfectly proportional to signal strength, a problem solved by using only the straight portion of this graph by biasing the signal*

This problem is overcome by biasing the tapehead, meaning some magnetizing current is allowed to flow through the tapehead irrespective of signal currents. Early recorders, and some current models designed for speech only, used either permanent magnets or a steady current through the tapehead to accomplish bias. Such systems are not suitable for recordings of reasonable quality, however, because distortion level is still high and there is a large amount of noise signal put on to the tape.

The system used for quality recordings is supersonic bias, in which the bias is a signal rather than a steady current, varying at a rate which makes it too high-pitched to hear (the original and correct meaning of supersonic). The amount of bias signal that is used is critical, and the setting is always a compromise affected by type of tape used and quality of reproduction. Excessive bias reduces distortion satisfactorily, but it also reduces the amount of treble which can be recorded, creating a muffled effect. Too little bias, in contrast, allows better reproduction of higher notes, at the expense of greater distortion. Changing from one type of tape to another requires bias settings to be changed and this is done automatically by many types of recorders, which detect the tape as either plain ferric oxide, chrome or metal types. Detection is done by using a system of hole positions cut into the cassette, but a few cassette recorders are in addition capable of carrying out a self-adjustment audit of a tape – recording and replaying a set of test signals and altering the bias to an optimum value.

Bias is not so critical for recorders used purely for speech, so the only cause for concern over bias in recorders forming part of the language laboratory equipment is if bias failed altogether. A factor closely connected with bias is the amount of signal current in the recording heads. Early recorders required the operator to monitor this level of signal on a meter and to adjust the input volume control, so the peak permitted level was not exceeded and the average level of signal did not fall below a reasonable minimum. This form of adjustment is still required for good-quality recording from radio and similar sources but, when the signal source is a microphone it is never used other than for high-grade professional equipment. The sound level at the input to a microphone is never predictable and few users of recorders have the luxury of a recording engineer to look after machinery.

Equipment designed mainly for microphone recording uses entirely automatic volume level setting, obviating the need to keep making adjustments, and requiring only some common-sense in the use of the microphone. Many machines also use automatic recording levels for other input signals, but these can lead to rather unsatisfactory recording of music, with loud notes followed by a considerable reduction in volume, and long quiet passages transformed as volume increases steadily throughout. The only noticeable effect of automatic recording levels on speech is the quite desirable effect of suppressing background noise when speech starts. This is because in the absence of speech close to the microphone, amplification is at its maximum level and is reduced when speech starts, making it much less sensitive to other noises.

For high-quality reproduction, equipment should use manual volume setting on record, monitored by a light-display rather than a meter and using a noise reduction system such as *Dolby C* or *dbx*. The *Dolby B* system which is used on pre-recorded cassettes provides a moderate amount of noise-reduction, but the use of *Dolby C*, *dbx*, or the newer *Dolby S* allows cassettes to be replayed with virtually no added noise so the signal is almost as noiseless as the original. Equipment of this calibre is justified only if amplifiers and loudspeakers are capable of taking advantage of it, and if it is intended for reproducing music to discerning ears. More recent developments in cassette technology include use of digital signals (like those on CD recordings) and use of new bias systems, but the existing technology is likely to endure for some considerable time because of the large amount of recorded material which exists in ordinary cassette form.

The recording head and tape drive system of a modern recorder is illustrated in Figure 4.3. Each cassette contains a spring-loaded pressure-pad which keeps the tape pressed against the record/replay head (a few systems use separate record and replay heads), while tape is moved past the head at a constant speed by the steady rotation of the capstan shaft as the pinchwheel presses the tape against the capstan. Tape is held against the erase head simply by tension on the tape at that point – contact between erase head and tape is much less critical than that between tape and record/replay head. Pressing the *play* key of the recorder forces the pinchwheel into position, using a spring to maintain correct pressure and moves the record/play head into contact with the tape. The take-up spool of the cassette is driven through a slipping clutch or belt, often by another electric motor, to keep tension on the tape fairly constant. Unlike old-style open-reel recorders, no threading of tape is needed for cassette machines, because tape path is determined within the cassette and placing a cassette into the recorder inserts the capstan.

Most machines use the same tape head for both recording and replay, and a few miniature speech-only machines use a permanent magnet for bias and erase, though it is much more usual to have a separate electromagnetic head for erase, which is activated only when recording tape. An important part of the record/replay head is the head gap; a tiny slit between two pieces of magnetic material, because the magnetizing effort is concentrated in this region. The head gap itself is invisible except under a high-power microscope, but its size and state of cleanliness has a huge effect on quality of recording and reproduction particularly when the speed of tape movement is slow, as it is for all cassette machines. It is particularly important to keep all metal, glass, ceramic or other hard materials away from tape heads, because a damaged head gap cannot be repaired, the complete head must be replaced.Magnets and magnetic materials should be avoided also, because anything which magnetizes the head permanently leads to all subsequent recordings having a high noise level, and even to

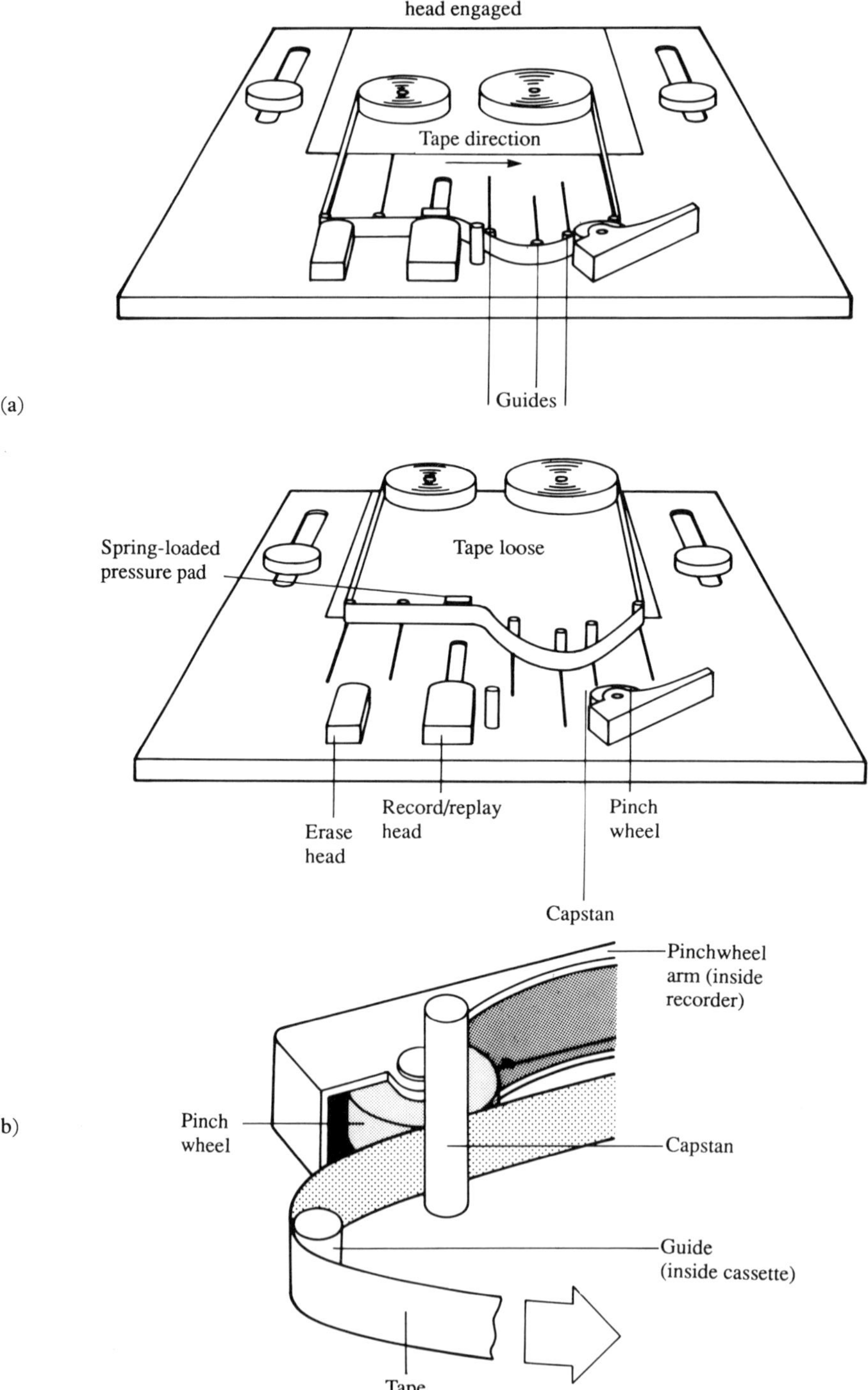

Figure 4.3 *Recording head and drive mechanism (a) of a modern recorder. The tape is driven by the revolving capstan, clamped by the pinch-wheel (b)*

some loss of signal from recorded tapes. Modern tape heads are made from materials which do not magnetize permanently to any great extent, but they should be kept well away from magnets in any case, as should recorded cassettes. A typical head gap dimension is 10μ (that is, 10 millionths of a metre – around 0.0004"). For a high-quality machine, a head gap of 1μ (0.00004") or less is used.

Erasing action consists of passing the tape over a head whose signal is the supersonic bias signal, but at a much greater amplitude. The erase head uses a wider gap (still not necessarily visible) and on some machines the amount of erase signal can be reduced or cut out, so that superimposed recordings are possible. This is not a particularly satisfactory way of achieving superimposition effects, and if these are often needed (for sound plays, for example), it is better to use a mixer and two recorders to achieve the superimposition.

The main enemy, both in recording and playback, is hum. Hum is the sound of the alternating current mains, and it can be picked up on any cable or connector unless care is taken. Construction of microphones, cables and connectors should reduce hum to insignificant levels unless very high volume control settings are used and if hum is obtrusive it is because of a fault in the system; it is certainly not an acceptable accompaniment to any recording. Given a reasonable standard of design, hum usually results from poorly-fitted connections or absence of a good earth lead on equipment; possibly use of different earth connections – Figure 4.4 illustrates good and bad earthing practices. Where mains wiring is up to date, the earth lead of a three-pin plug should provide as good an earth as is needed for hum reduction and, if not, it is advisable to have mains sockets checked – a socket can sometimes be used for years without having an adequate earth connection and it is better to find this out by tracing a hum problem rather than someone suffering a fatal shock if faulty equipment is connected to the faulty socket.

User guide – playback only

Users of tape equipment are divided into language laboratory users and the rest. The rest are considered first, subdivided into those who use recorders for replaying only and those who carry out recording as well, because requirements are not identical. In

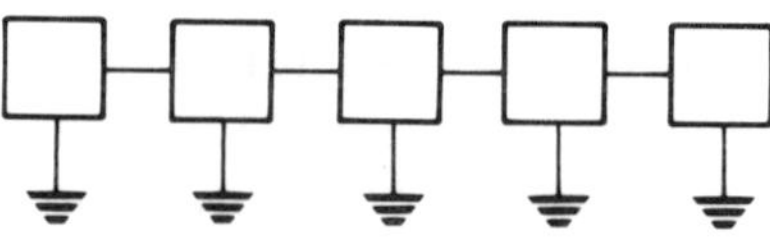

(a) Separate earths – hum is likely

(b) Common earth – hum reduced

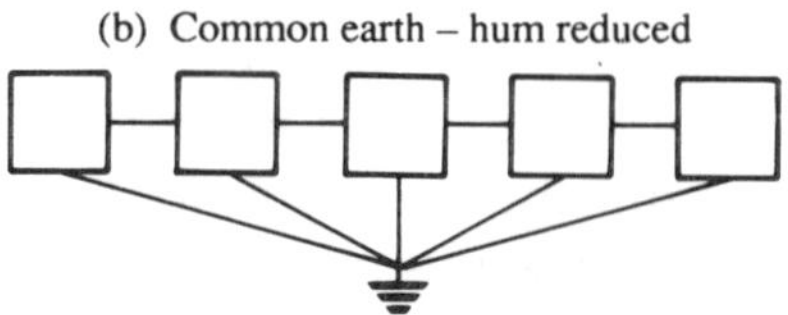

Figure 4.4 *Earthing – where several different units are used, one common earth is preferable to a set of different earth points*

addition, the form of cassette recording equipment these different groups use is different.

Users of replay-only equipment are probably primarily interested in reproduction of music or speech whose content is important (such as recorded plays where characters are distinguished by voice). This calls for a machine capable of good quality reproduction; such a machine will use a noise-reduction system of at least *Dolby B* standard, possibly offering *Dolby C* or *dbx* in addition. Note, however, pre-recorded material which features noise reduction always uses *Dolby-B;* only recordings made by a user or by other users with similar equipment feature superior *Dolby C* or *dbx* methods. The reason for use of only *Dolby B* on pre-recorded material is that only *Dolby B* permits replay of noise-reduced cassettes on machines with no form of noise-reduction circuits, so *Dolby B* recorded cassettes are compatible with all machines. To gain any advantage from a noise-reduction system, the same system has to be used when replaying the tape as was used when recording. Noise-reduced recordings made using systems other than *Dolby B* sound very peculiar when replayed: when a *Dolby B* cassette is replayed without any form of noise-reduction system the only noticeable effect is a rather shrill treble, which is dealt with by using the treble-cut control on the replay machine.

It may seem rather unjust that recordings of high-quality music must be made with a noise-reduction system less favourable for high-quality reproduction, simply so they can be used on small portable cassette units, but that's the commercial world. The arrival of compact disc and digital tape systems makes it unlikely that more advanced noise-reduction systems will be accepted in commercially produced cassettes. It is possible, however, compatible cassettes will be developed which can be played in the ordinary way on existing machine, and also as digital cassettes on suitable equipment.

It is not sufficient to make use of a good cassette machine with a suitable noise-reduction system; remaining equipment also needs to be geared to high-quality reproduction. Loudspeakers which are adequate for use with movie projectors are decidedly not suited to this type of work. Though suitable loudspeakers are more expensive, they are not nearly so expensive as some of the educational equipment bought in haste (but never used) as the end of the financial year approaches. They are also likely to have a considerably longer life if they are kept away from the more accident-prone members of staff. Note that disco speakers are not suitable for high-quality reproduction. Besides being unused to the sound of music, they are generally designed to emphasize bass thumps at the expense of the rest of the spectrum of musical notes and give very unbalanced results with musical sources.

Recordings are almost invariably in stereo, so loudspeakers should be arranged to make the best of stereo replay, with its very considerable advantages of sense of position, which is particularly useful for recordings of stage plays and orchestral music. Figure 4.5 shows arrangements of speakers in rooms, using either a long wall or a short wall. By siting the speakers on a short wall the area in which the stereo effect is clearly heard is larger, and also bass response of loudspeakers tends to be better. Speakers should not be placed against a wall, but a couple of feet out, avoiding corners of the room. Separation of loudspeakers is also important because a very poor effect is obtained if the loudspeakers are too close together. Suitable loudspeakers are heavy even if they are not large, and should be positioned beforehand if possible.

The most difficult requirement to meet for high-quality sound reproduction is a silent background. This is particularly important for music, because the contrast in sound intensity between, say, full orchestra and pianissimo solo flute, is enormous. If background noise swamps the softest notes it is impossible to obtain the correct ratio of sound levels, because the ear responds differently to a range of pitches of music when volume is altered from its original level.

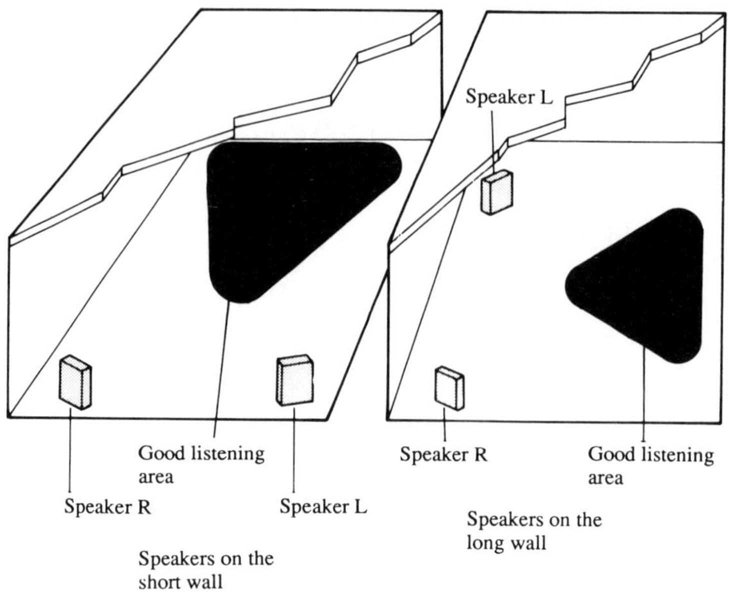

Figure 4.5 *Stereo effect is better in rooms where loudspeakers are situated on shorter walls, as this provides a larger space in which the stereo illusion is present*

Replaying should start by ensuring the volume control of the amplifier is at its minimum setting, the balance control in its midway position, and any bass and treble controls in their mid positions, (neither boosting nor cutting bass or treble). If the room is known to require particular settings for bass and treble controls, these positions should be marked, and settings made in advance. The amplifier can then be switched on, followed by the cassette recorder. If switching the amplifier on produces a loud thump from the loudspeakers, the technician should be told – a small noise is acceptable but a loud one is not and points to an amplifier fault.

The cassette can now be inserted, and settings of the noise-reduction control checked – adjustment for tape type is usually automatic, but if not then the correct tape type is manually selected. Although most pre-recorded tapes use conventional iron-oxide tapes, several are on chrome tape while a few recordings exist on metal powder tapes. The play key should be depressed, and the volume control advanced. When the indicator of the cassette recorder shows replay has started, the volume control can be taken to its normal playing position. This avoids the problem of starting with a volume level too high, but if this is deliberately intended (perhaps to impress with Aaron Copeland's Fanfare) the volume control setting should be marked beforehand and set to the marked level before starting the tape.

During playback it may be necessary to make adjustments to volume and bass and treble controls, but this should be done only after listening from a position in the room corresponding to the 'average' listener, because listening from a position too close to the loudspeakers does not provide more than an impression of how the sound is received by the bulk of the audience.

Care of cassettes is another important facet of using recorded sound. Cassettes should never be stored partly-wound, and should be fully wound or rewound to one end of the tape after use. The piece of tape exposed in a cassette is vulnerable to contamination by dust, sticky fingers, smoke and a score of other pollutants, and this must never be allowed to happen to a piece of recorded tape. Leader and trailer sections of tape (at each end) are usually uncoated and of tougher material. This is necessary, because these sections take the strain when a tape is fastwound to a halt in either direction, and they can also cope with being exposed to atmospheres that are less than perfect. In addition, magnetic tape suffers from the effect called print-through, in which a strongly recorded piece of tape copies on to the pieces of tape closest to it on a cassette reel. This can cause the effect called 'pre-echo', in which a faint copy of a loud sound occurs just *before* the loud passage, the time interval being the time needed to turn the cassette reel by one revolution. Print-through is greatly reduced if cassettes are kept in cool conditions and regularly wound from one end to the other, ensuring that contact between pieces of the tape is changed from time to time. Print-through is particularly noticeable on thinner tapes, so C120 tape should be avoided.

For high-quality work, choice of cassettes is important. Few recordings of significance are likely to make use of C120 tapes, which play for one hour in each direction, two hours total playing time. Tape in these cassettes is very thin because a greater length (one third longer than a C90) must be accommodated within the same space on the standard-sized cassette. This makes the tape much more liable to breakage and to print-through. Most recordings are available on C90 or shorter tapes – commercial recordings are on tapes usually just long enough to take the main item or to allow for, say, four movements to be split into two each side. Tapes made from live performances or from radio broadcasts (under copyright regulations) are liable to be less orderly, often with one side recorded to a larger extent than the other and using C90 tapes.

Cassettes should be kept in their casing when not in use, with clearly labelled contents, length of performance on each side, and date of recording if known. Any cassette which jams at any time should be handed to the AV technician with a request to re-house the tape. When a tape jams it is foolish to expect it will release itself and the general rule is that jamming is a warning of worse to come. The best treatment is to wind the tape into another cassette. This is a task for an experienced technician.

User guide – record and playback

The second group of cassette users consists of those who use the machines for both recording and replay, often with the emphasis on topicality and interest rather than high-quality of sound reproduction. The main sources of material are microphones and radio broadcasts (subject to copyright) and users in educational establishments should note there is no overall copyright agreement for pre-recorded material on LP records and compact discs and other tapes as there is with BBC broadcasts. Permission to copy tracks from an LP, compact disk or tape must be given in writing by the owners of the copyright. This is often comparatively easy to obtain when the source is an old LP or 78 recording which is no longer available; some old 78 recordings are out of copyright in any case, or the copyright cannot be traced. It is much more difficult to obtain permission for copying CD or tape, because such recordings are usually current ones, but in some cases, copyright waivers can be negotiated by the local educational authority, particularly if there is agreement about erasing copies after use. If this is done, it is important to log all copies and to be able to account for them. Copyright prosecutions for tape copying are virtually unknown unless some commercial gain has been involved, but this does not make such copying legal.

Recording from radio is a daily routine, in primary education in particular, and machines used are generally fitted with time-switches. This, curiously enough, is rare in domestic audio equipment. We would not contemplate buying a video recorder which did not include a timer, but it is very difficult to find audio equipment other than specialized designs for educational use which incorporates such a convenience. This is mainly because high-quality audio equipment is bought in sections, and incorporating a timer into a recorder is of little use unless it also switches on a tuner and an amplifier and ensures the tuner is switched to the desired broadcast. Recording of high-quality material must be done manually because the recording volume may need to be altered during the recording, though a few recorders which use the *dbx* noise reduction system are very tolerant of high input volumes and can often be left on a fixed setting.

Recording is straightforward when setting of recorded volume is automatic, and such recordings may have to be made by the user or, in the case of primary schools, by the school secretary. Points to remember are:

- Leader and trailer ends of tape cannot be recorded on. Always wind a tape on, using a ball-pen or pencil with a hexagonal body as a shaft to drive the reel of the cassette, until the brown recording tape appears in front of the pressure pad.
- Some timers use the 24-hour clock ('military time' in the USA) and others use the AM-PM convention. Be careful not to confuse the two.
- Radio broadcasts aimed at educational users generally run to time, unlike televised material.
- Always check a tape before recording in case it contains a cherished recording left unlabelled.
- Recordings to be kept (made from other sound sources) should have the recording protection tab (Figure 4.6) broken off. If a recording has to be made subsequently, the space can be covered with sticky tape.

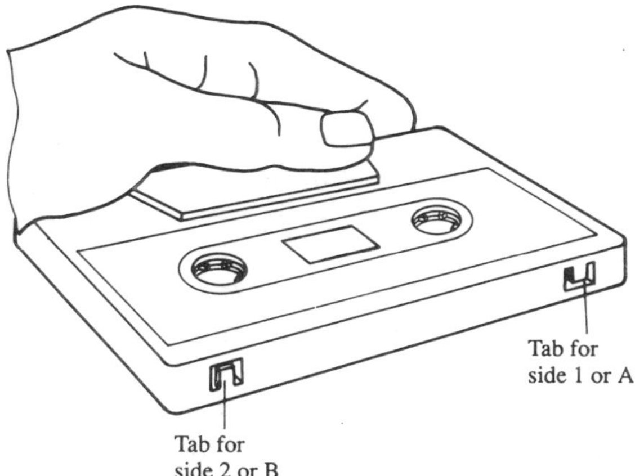

Figure 4.6 *Record-protection tabs of a cassette – break off to prevent inadvertent re-use of the tape. If the tape is to be re-used, the hole can be covered by sticky tape*

Microphone recording demands rather different techniques, and is virtually always carried out by someone other than the AV technician. At one time microphones fitted to cassette recorders were of very low quality, and a better microphone had to be used. This situation has changed, and the developing technology of microphones has resulted in cassette recorders, even in a low price bracket, with in-built microphones of reasonable quality. Such microphones are, however, generally incorporated into the recorder casing and this is far from ideal unless the machine is being used as a dictaphone.

A built-in microphone is liable to pick up mechanical noises of the cassette recorder itself and the surface on which it is placed, and is seldom in an optimum position for recording speech or music. A microphone is not a human ear and does not respond like a human ear; in particular it does not have the ability to concentrate on a wanted sound to the exclusion of others. In addition, recorders designed to make use of a microphone always incorporate automatic volume control, which attempts to ensure that signal amplitude on the tape is always as high as can be recorded.

The effects of this are familiar to anyone who has suffered listening to a series of amateur recordings. The cassette recorder with its built-in microphone is left on a table at the back of a hall, and the performers on the stage make their entrances, their speeches and their exits. As the intensity of a sound wave diminishes with increasing distance (each doubling of distance causes intensity to drop to a quarter of its previous value), the recorder, in such circumstances, operates at maximum sensitivity. In addition, the microphone, unlike the human ear, is particularly sensitive to low-pitched sounds and the result is a tape that faithfully records each slammed door at the other end of the building, each distant Boeing coming in to land, road traffic, rail noise and, somewhere in the background, a hollow-sounding human voice.

Hollowness of voices is due to echoes, which exist in every room. The ear tends to ignore these unless they are very prominent, but as far as the microphone is concerned, many of these echoes are louder than the direct sound from the stage and are therefore recorded at a higher level. This makes all such recordings sound very 'boomy', as if recorded in a dungeon. There is a notable lack of treble caused by the echoes, as treble notes are more readily absorbed by surfaces and consequently do not contribute so much to echoes. If all you are concerned with is to prove that someone has fluffed a few lines this might be adequate, but it is not very difficult to improve on such a standard of recording.

The main improvement is to keep the microphone closer to the source of sound. In this way, a much greater signal volume ensures the recorder works at a lower sensitivity level and this, in turn, rules out background noises. It also makes the direct sound of a much greater amplitude than the echo, giving more prominence to treble notes so the sound is 'fresher', less booming and considerably better to listen to. Even when using a recorder's built-in microphone, it is usually possible for this to work satisfactorily for a solo performer, but for staged plays or for anything more than a trio of players an external microphone is much more satisfactory.

Internal microphones of cassette recorders are generally omnidirectional, meaning their sensitivity does not depend on being pointed in any particular direction. This is counteracted to some extent by the way the microphone is located in the recorder casing, but in general these microphones are, unlike the human ear, incapable of emphasizing pickup of sound by being pointed to the sound source. It is possible to buy directional microphones which as the name suggests, respond best to sounds within a narrow angle of a line drawn straight from the axis of the microphone. Without getting into technical details, the way these two types of microphone perform is summarized in diagrams of the form shown in Figure 4.7, showing contour lines of equal sensitivity. A perfectly omnidirectional microphone is characterized by circular contours, a perfectly directional one by a straight line (equally sensitive to all sounds coming along the axis from any distance). No microphone is ever perfectly one or the other type, but it is not difficult to buy microphones which are predominantly of one class or the other. Most cassette recorders provide for the connection of one external microphone, and making this connection will automatically cut out the internal microphone.

Directional microphones are much easier to work with, and are preferred for anything except the type of performance where performers are located in a circle round the microphone. If a single directional microphone is to be used, it should not be too directional, particularly for a performance recorded from a stage. Use of two or more directional microphones, or combination of a directional microphone and an omnidirectional type can provide, with practice, much better results. Stereo recording is best carried out using crossed directional microphones as illustrated in Figure 4.8. When more than one microphone is in use, signals have to be amplified and mixed prior to recording, and this is a task for an

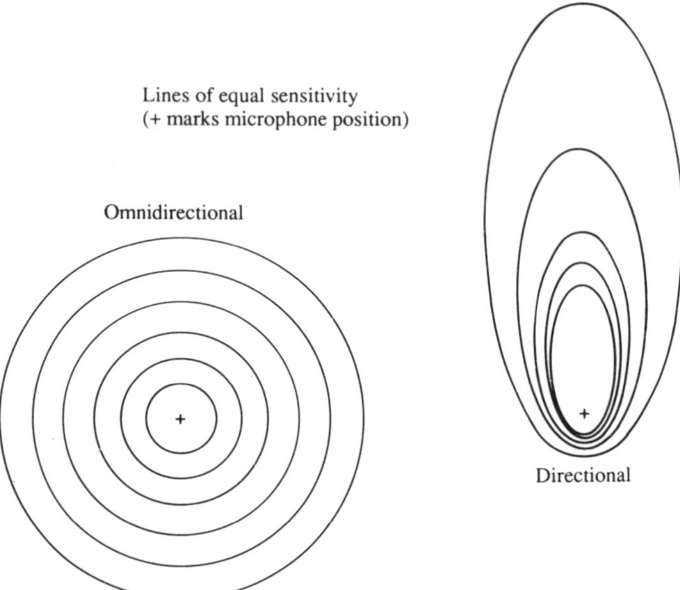

Figure 4.7   *Microphone directionality, shown as contour lines of equal sensitivity. For the omnidirectional microphone, these lines are circles, but the directional type is much more sensitive in one direction than in any other*

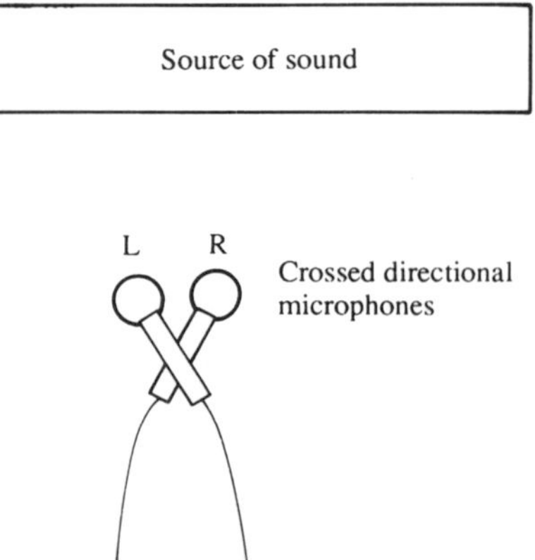

Figure 4.8   *The classic arrangement for stereo recording (dating back to 1936), using two crossed directional microphones*

experienced AV technician or someone with an interest in and experience of sound mixing. Like so much of AV work, this calls for a mixture of technical knowledge, artistic appreciation, and judgement.

Interviews, by contrast, are easier to record, because it is enough to ensure that the microphone is kept at a constant distance from the subject and voice level is fairly constant. Once again, though, use of a directional microphone makes the task much easier, providing the user remembers to point the microphone in the correct direction, and to change this direction when the sound source changes. The most difficult interview recordings are those made with a subject who is moving ('tell me about your Marathon successes, Mr Jones') or one who considerably alters the volume of his/her voice at intervals ('I'll sing you the one about the pig in the five-acre field'). For such eventualities, an omnidirectional microphone can be better suited, altering the distance between microphone and subject to cope with sudden changes of volume.

Language laboratory users

Language laboratory use of recorders is a much more specialized aspect of cassette recording than other uses described here, and users must be thoroughly trained in the system used. Modern education lays a lot of emphasis on understanding principles (we practice education, not training), so it is not unreasonable to expect teachers who use the language laboratory to understand how the system works (and as most teachers drive cars, should they not understand the principles of the internal-combustion engine, or is emphasis on education only for the consumers?). As any general description tends to be vague, the following refers to the Cybervox Series 400, a particularly successful system.

The essence of a language laboratory is that each student must learn independently, but with central guidance from an instructor. Early designs of language laboratories were built round open-reel tape machines, requiring a lot of maintenance and easily wrecked by careless use. In addition, contact between teacher and student was very primitive, and it was difficult for the teacher to monitor work adequately. This led to considerable disillusionment with the whole principle, often to total abandonment of a language laboratory when machines needed servicing, or to virtual abandonment of language teaching. Modern computer-based systems have transformed language laboratory design out of all recognition, but have made necessary security of premises and equipment even more important than it was in the past.

In a modern system the teacher has visual information presented on a screen. This screen typically shows messages from individual students (help wanted or cassette not inserted), student information such as identification number, group identification (no use having the GCSE French cassettes in place for the *Japanese for Businessmen* course), remote recorder in contact, counter number of remote recorder and current action, information about the program source and remote control action, and a reminder of the actions of control keys. This places less of a load on the memory, and avoids the need to search frantically through the manual to find what key number 5 does. Such a console display provides manual instructions as and when required, and allows the teacher to choose language programs, tests and other actions with a reminder showing on screen. Information received from students is displayed clearly, and there is a permanently displayed record of student information (no need to peer in each booth to fill in a register).

Such a system provides clear instructions for the less-experienced user, such as 'place tape in recorder', and also shows if a student is experiencing mechanical problems (no cassette in student's recorder, recorder not connected, tape stopped), and the state of each student's recorder, including the tape counter number.

A typical console is illustrated in Figure 4.9, showing screen, student keys, action keys and soft keys. Student keys allow access to each student's booth and recorder system; so pressing one of these keys allows the teacher to monitor work of one particular student. Action keys are used to call up information, such as key functions, or for over-riding actions such as calling all students, controlling all remote recorders or selective intercom discussion with one student selected from the student keys, with control over the recorder that is being used by that student. The most important point is that the screen shows instructions to the teacher which, if followed, avoids misuse of the system, guides selection of actions, reminds him/her of system capabilities and reduces the load on memory, allowing the user to concentrate on teaching.

Computer control makes it possible to run in an *autolesson* mode, in which the following set of actions is carried out without intervention:

- Students recorders are rewound, then set for recording.
- Master recorder starts playback.
- Teacher's display screen monitors work of all students.

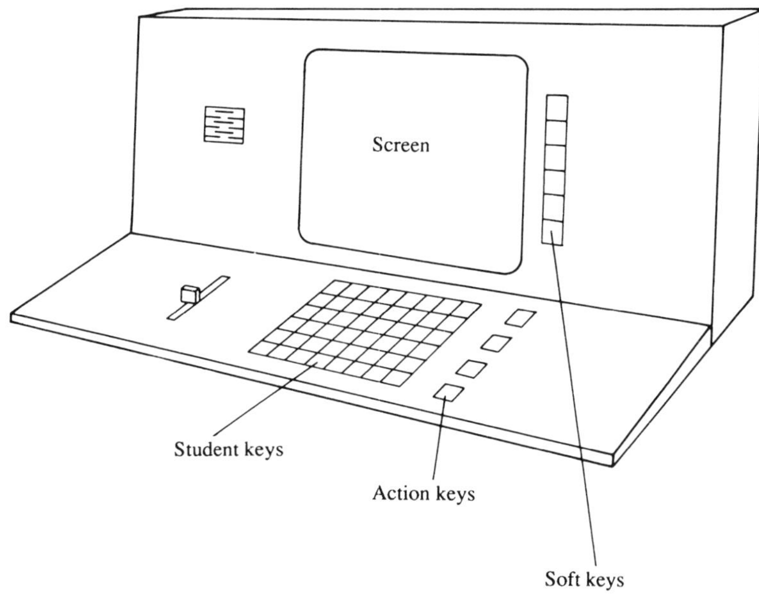

Figure 4.9 *A typical console for a modern computer-controlled language laboratory (courtesy Cybervox)*

- Master recorder stops at end of program (or a 30 second gap).
- Student recorders are rewound.
- Student recorders are returned to control of students.

Technical maintenance

Maintenance work by the AV technician can be divided into the groups of general use and language laboratory. At one time, the language laboratory would take up much of the time of the AV technician, but adoption of modern computer-controlled systems has greatly reduced need for maintenance. Where old systems are in use, however, the AV technician may have to work in the language laboratory whenever a class is not in progress, and *may* have to work on unoccupied booths even when a class is present.

Dealing with more general uses of cassette recorders first, maintenance work mainly concerns cleaning of recording head, capstan and pinch wheel, and the task of cassette salvage. Even when recorders are intensively used, head cleaning is not normally necessary at frequent intervals, and in some cases can be done once a year, though a better system is to carry out this action once per term. On most recorders, head cleaning does not require any dismantling because heads are accessible when the recorder is put into play mode with no cassette in place. If there is a cassette-sensing switch which prevents the machine from being switched on without a cassette, this can usually be pushed down with a matchstick or similar tool to allow the play key to be depressed.

Build-up of material, mainly magnetic metal oxide used as a magnetic material for tapes, on heads causes sound quality to deteriorate noticeably. The difference is most appreciable when response to high-pitched notes is checked, because the main effect of deposit on the head gap is to reduce treble response. The abrasive nature of tape also wears heads away, if heads are made from metal, and this also causes a (permanent) reduction of treble response. Many recorders are fitted with ferrite heads, using a glassy material which is not abraded by the tape, and life of such heads is likely to be very long. If cleaning does not improve the response of a machine which uses metal heads, head replacement is probably required. This can be carried out with comparatively simple equipment, but not every AV department has the facilities to align the replacement heads effectively, and it may be preferable to send the machine(s) for specialist attention. Many local educational authorities maintain central workshops for such purposes.

Whenever tape heads are cleaned they should also be degaussed (demagnetized or defluxed) using a proprietary head defluxer. This device resembles an electric toothbrush and is effectively a form of tapehead itself covered in plastic. The plastic coating prevents metal of the defluxer touching tapeheads, and the defluxing action consists of switching the defluxer on, placing the tip on to the record-replay head of the recorder, and then slowly (over a period of about a minute) drawing the tip away from the head until it is several inches away, when it can be more rapidly withdrawn and switched off. An alternating current through the defluxer magnetizes the head alternately in each direction some 50 times per second. The action of slowly drawing the defluxer away gradually reduces the magnetization until it is negligible. This removes any trace of permanent magnetization in heads which might cause noisy recordings.

The effects of deposits from the tape on to the capstan and pinch wheel are to cause the tape speed to increase and decrease alternately at a high speed if the lump is on the capstan, slower if it affects the pinchwheel. The effect on sound is to make steady musical notes sound tremulous, and the effect is known as 'wow' if it occurs slowly, and 'flutter' if it is fast. When a machine is used mainly for speech recording, wow and flutter are not usually noticeable until music is played and a user may blame the tape rather than the machine. Tape is generally responsible only if the recording has been made on a machine with these faults.

Cleaning should be carried out using a felt pad (Figure 4.10) and isopropanol (iso-propyl alcohol). Kits containing these materials

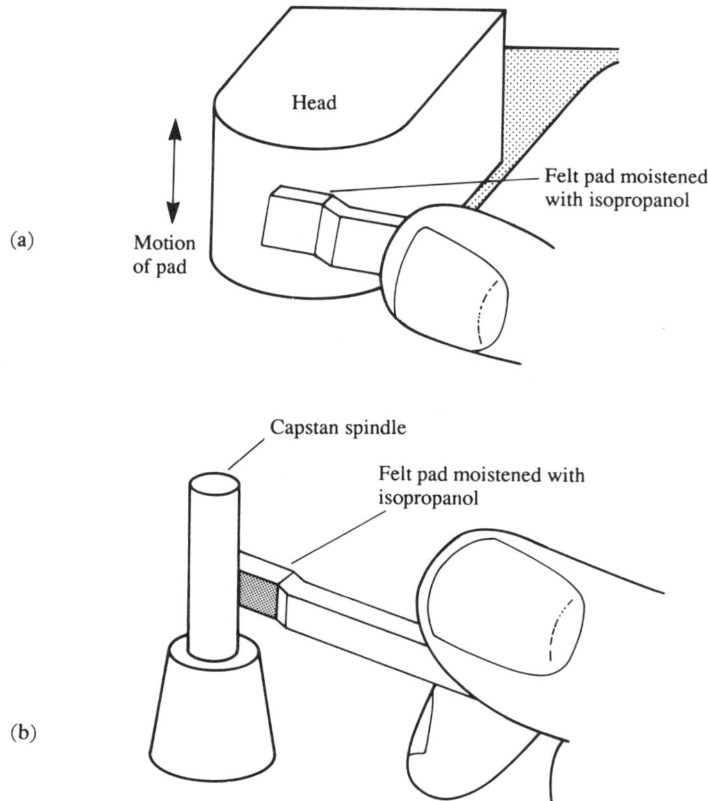

Figure 4.10 *Cleaning a cassette recorder. Cleaning kits can be bought containing isopropanol and felt pads. Care must be taken not to scratch the tapehead surfaces (a) avoiding all metal or other hard objects. The capstan spindle (b) is also vulnerable and must not be bent*

can be bought in any shop specializing in audio equipment and in many chain stores. Do not use the 'cleaning cassettes' widely available, because they are not particularly useful for heavily-used recorders. Cotton buds and methylated spirit should also be ruled out on two counts. One is that cotton wool can leave strands inside the recorder. Second is that methylated spirits usually leave a sticky residue on tapeheads which causes immense trouble. Isopropanol is the only suitable solvent that is widely available, and its flammability (hence risk of fire) is lower than that of absolute alcohol (ethanol). The felt pad should be checked to make sure it does not pick up any slivers of metal or glass which could scratch the heads, and nothing hard should ever be allowed near the head area. A scratch on a head can act as a secondary gap, totally ruining head performance.

The felt pad is moistened with the isopropanol, not soaked and certainly not dripping. It is rubbed firmly over the heads, particularly the read/write head, then a dry pad is rubbed over the head to remove any residual film of solvent. Heads should be left exposed in an airstream to dry thoroughly before the machine is tested and put back into use. A note should be made of the cleaning operation so you are sure you are not cleaning the heads of the same machine more than once in a session.

After heads have been cleaned, the same cleaning treatment is used on capstan and pinchwheel. By using the recorder in play mode, the capstan spins making it easier to ensure even cleaning. Avoid excess pressure on the capstan, because if a capstan is bent by excessive force there is no possibility of repairing it, and the whole unit (often the motor as well as the capstan) has to be replaced. The pinchwheel is not so easy to clean, and it is not usually possible to remove all traces of deposit, though it should be possible to leave only a uniform film of brown material.

Cassettes handed to the AV technician for salvage are invariably in a sorry state, with the tape jamming, broken or even wrapped tightly round the capstan of the recorder. Broken tape can usually be spliced, and the only difficult part of the exercise is getting the broken ends out of the cassette, for repair in a tape splicer (Figure 4.11). If the ends cannot be fished out, the cassette must be opened, either by unscrewing it if it is of the screw-retained construction, or by cutting if it is of welded construction. Tape ends can then be located, and reels put into a salvage cassette (Figure 4.12) while the ends are spliced. Use only tape sold for splicing purposes, because ordinary transparent tape uses adhesives which spread in time and can bond several layers of tape together during storage in a warm place. If a salvage cassette is used, the cover is screwed on when the tape has been spliced, and the tape is checked. The splice is always noticeable, but no worse than a small scratch on an LP disc. Note that splicers and adhesive

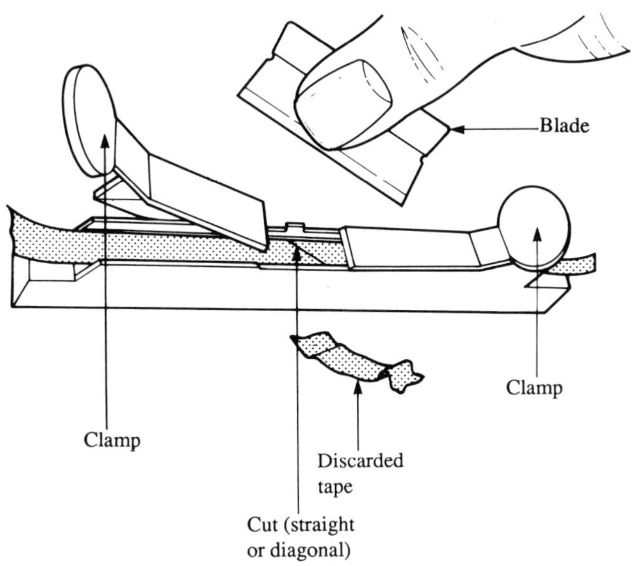

Figure 4.11   *Repairing tape by splicing, using a commercial repair kit such as the BIB. The tape is cut on each side of the break to present a straight join, and is joined by splicing tape, which uses a non-spreading adhesive*

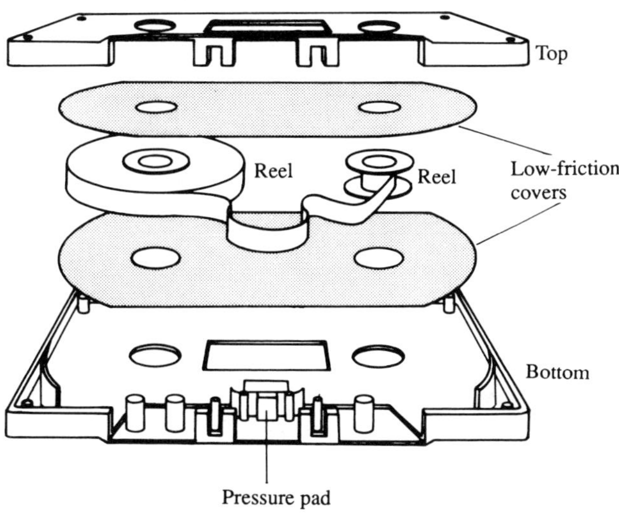

Figure 4.12   *A salvage cassette can be used to hold a pair of reels and tape from a damaged or jamming cassette — remember to position the low-friction plastic sheets correctly in each side, otherwise the rehoused cassette will stick*

tape intended for open-reel tape (0.25″ wide) cannot be used satisfactorily to splice narrower cassette tape (0.125″ wide).

A jammed cassette can often be wound in one direction or the other by hand, using a hexagonal ballpoint pen or pencil. If the tape is not badly jammed, this may release it enough to allow it to be fast-wound, first in one direction then in the other. This may cure the jamming if it is caused by bad layering of the tape which is usually caused, in turn, by frequent stopping and starting. If jamming is not amenable to this treatment, the cassette should always have its reels transferred to a salvage cassette and if there is any sign of damage to the reels, tape should be wound on to the reels from the salvage cassette. An old open-reel tape recorder can usually be adapted to carry out this rewinding more easily; a useful alternative is a hand-winder. A jammed cassette requiring salvage in this way is usually caused by a warped casing, and no amount of rewinding eases the tape. The fault should be investigated, because it may mean that a recorder is overheating in use.

If a tape is wound around the capstan and pinchwheel, it can sometimes be freed by first disconnecting the power supply then trying to turn the capstan and pinchwheel backwards by hand, pulling the cassette out as the tape releases. If the jam cannot be treated in this way, the mechanism must be dismantled, though if the tape is not important an option is simply to cut the tape away, though great care is needed if scissors are taken near heads and capstan.

The normal action of Murphy's law, though, dictates that any tape suffering this fate is the only copy of a particularly irreplacable recording, so every effort must be made to get the tape out intact, even if this means extensive dismantling of the mechanism. Once the tape is removed, a matter for patience rather than skill, it can be wound back into the cassette, and the cassette fast-wound in each direction then copied keeping a watch on the tape in case the jam should recur. If the original recorder is dismantled to remove a tape, it should be carefully reassembled, tested and then put back into service. Tape spillage and jamming is often caused by a broken drive to the takeup spool of the recorder, and this must be checked.

One important point concerning loudspeakers arises with stereo replay. Connections to loudspeakers are polarized so they can be inserted only one way round. This avoids out-of-phase connection of speakers, which has the effect of markedly reducing bass response on any replayed recording.

Language laboratory maintenance

The main item of language laboratory maintenance is periodic cleaning of equipment, particularly student master recorders, paying particular attention to record and erase heads, capstan and pinchwheels on each machine. In addition, any reported defects should be investigated, and repaired (if possible) or a service call requested. Following this routine, the rest of the equipment should be checked, using a sequence such as the following:

- Check program sources, using the teacher's direct listening option to monitor sound from whatever sources (such as open-reel recorder, cassette master recorder, radio, disc) are in use.
- Check that each student's booth recorder responds to commands from the central console (listen, fast rewind, stop, record actions).
- Using a set of blank cassettes reserved for testing purposes, record a short extract from a sound source on to all of the student recorders. Check that each machine plays this back correctly.
- Repeat the recording test at the highest recording speed available.
- Rewind the tapes and record to all machines from a blank input (a radio which is switched off, for example). Monitor tapes to ensure that no trace of a previous recording remains.
- Log any faults found to report to the service engineer, or for further action if servicing of equipment is internal.

Hints/tips

- Never try to economize on tape. Cheap tapes of unknown origin often jam or break.
- Do not assume all tapes use the same length of leader and trailer.
- All tapes used as master tapes should be record-protected by breaking off the tab.
- Very distorted recordings (check the replay on more than one machine) can be caused by bias failure. If an oscilloscope is available, check the presence of a bias waveform at the heads.
- Mechanical faults are much more common than electronic faults, but the electronic system is more often blamed.

- Crackling sounds when a volume control is operated are caused by dirt and grease within the volume control potentiometer. A spray with WD40 or similar silicone oil can work wonders without even requiring the potentiometer to be opened, as there is usually a vent through which the oil can be squirted, using the fine-bore tube that is supplied.
- Faulty switch contacts are also a problem dealt with by using WD40. Spray a small amount on to the switch contacts and operate the switch several times.
- Remember that tapes should not remain in hot places for long periods, as heating greatly accelerates print-through and loss of recorded signal.
- Where there is a rapid turn-round in cassettes, a bulk-eraser is very useful for clearing cassettes quickly without the need to run the cassette through a recorder.

5

Video equipment

The introduction of video equipment, starting with the unadorned television receiver and progressing through video recorders to video cameras has been the most marked change in AV aids so far. Video, like computing, has benefitted from the accelerating pace of technology, so what could be accomplished rather indifferently and at great cost yesterday becomes easy to achieve at low cost today. This has the unfortunate side effect of penalizing the pioneers, who pay enormous amounts for the first efforts in a new technology only to find it becomes out of date after a year or so. It is not easy to determine the optimum time to buy equipment, and experience has often proved that people in the positions to make decisions are not always the best equipped to do so.

One underlying theme greatly stressed by companies which specialize in AV hardware is that such equipment must be made more durable than domestic grade equipment, hence much more expensive. Sometimes this no longer applies – the school video recorder probably receives no more wear and tear than the domestic version nowadays. There is, however, a case for differentiating between professional equipment intended for video studios feeding material for broadcasting and simpler domestic equipment, but use of professional broadcast equipment in schools, colleges and business is not justified.

Such equipment is designed to be used by professionals who understand it thoroughly, and to be serviced at frequent intervals. Neither of these conditions can be applied here – the costs of running just one professional grade outside-broadcast camera and recorder with appropriate staff would account for the salaries of most of the staff in a medium-sized school.

Mistakes of this type in the past have led to unusable equipment being accumulated, becoming out of date almost as soon as it was purchased and which, for lack of funds, cannot be repaired if anything goes wrong. By using domestic type of equipment, it is possible to keep costs down, in particular repair costs. In addition, if a domestic-grade video recorder becomes troublesome, its tapes can be played on any other domestic recorder, and most repairers are happy to lend a machine while another is in for repair. Exotic equipment usually requires specialist treatment, and if you prefer Lamborghinis to Minis you have to accept that the running costs, repair bills and driver skills are in a different league. In addition, such equipment often fetches only scrap prices when sold after a few years.

The video equipment in most frequent use is the video recorder and monitor, allowing broadcasts (subject to copyright agreement) to be recorded and then used at more convenient times, with the advantage that recorded programmes can be interrupted (for more explanation or possibly to express an opposite point of view) and resumed with no loss of content. Methods used are either the provision of trolleys (each carrying a recorder and monitor) which can be taken from room to room, or a cable ring with outlets in each room so signals can be transmitted from a central point. Cable systems sound attractive, but as a monitor still has to be taken to the room in which the video is to be used, reduction in corridor traffic is not noticeable and difficulties arise if several users want different programmes over the same cable at the same time. With the use of videocassette recorders operated individually, attractiveness of cable is greatly reduced.

Where cameras are in use for teaching reporting techniques or for demonstrating effects normally visible by only one person through a microscope, there is little chance in modern times of a technician being allocated permanently to camera control. This calls for small and easily operated video cameras unlike the immensely weighty beasts (themselves tiny in comparison to studio cameras) of the past. The quest for smaller cameras has led logically to the 8 mm tape format and this is now becoming established for educational and business use, slightly ahead of the domestic market in which the larger VHS (full-sized cassette) cameras still predominate. The main rival to the 8 mm format is the miniature VHS-C cassette, which can be played, using an adapter, on a standard VHS machine.

Security of equipment is particularly important. Light and portable items like cameras should be electronically tagged (preferably internally if possible), and larger equipment kept on trolleys should be secured to the trolleys – do not assume that simply screwing a recorder to a wooden surface will suffice. The greatest danger to portable equipment arises when no-one takes responsibility for returning it, leading to equipment being left in empty rooms; a certain invitation to the thief and the vandal.

Principles

Principles of video signalling and recording, though built on the experience of audio, are considerably more complex. In audio a sound wave is converted into an audio signal by the microphone; a conversion from one type of wave to another in which electrical voltage varies rather than air pressure. The range of wave repetition rates (frequencies) for audio is simply that which exists for sound sources, of the order of 100 to 10,000 repetitions of the wave per second, usually expressed as 100 hertz to 10 kilohertz. There is no such simple and natural conversion for a visual image.

The solution is the system called scanning, first devised by the pioneer Nipkow in 1884 and anticipated in its modern form by Campbell Swinton's suggestions in 1908. The principle is that a small sample of image is taken, and used to obtain an electrical signal proportional to brightness and colour. By taking samples in order and in rapid succession, a whole image can be converted into a set of signals in a time which is typically 1/25 second, so complete pictures repeat at this rate, very close to the 24 frames per second of the movie camera and projector.

Figure 5.1 shows how scanning can operate. The sampling area, shown here as a spot, moves across the image in a straight line scan which is slightly angled, and samples are taken at intervals corresponding to points where spots just overlap. At the end of a line the spot is rapidly moved back during the flyback, horizontally without taking any samples, and the slope of the scanned line ensures that the spot is now under the position that it formerly took at the start of the first line, by an amount approximately equal to spot diameter. The scanning across and flyback is now repeated until the spot position reaches the bottom of the image, when the spot will be returned to the top left hand corner.

Modern (since 1935) scanning systems use a system called interlace to overcome two conflicting requirements for a TV signal. One requirement is that the repetition rate for pictures should be fast, preferably faster than 25 times a second, to avoid flickering of the bright screen. The second is the range of frequencies in the signal (its bandwidth) should be as low as possible. In an interlaced

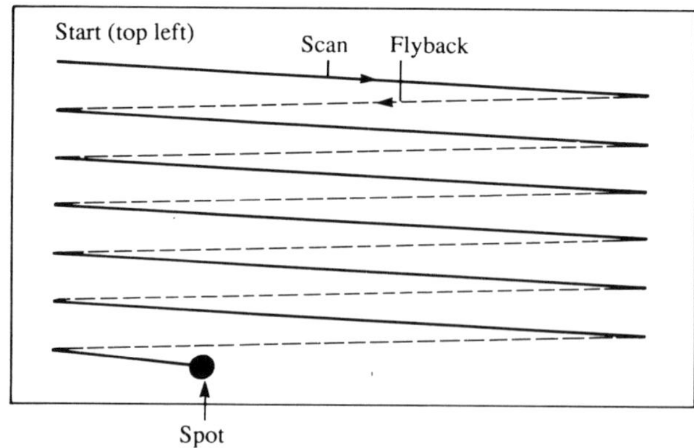

Figure 5.1   *Scanning as applied to television. The sampling spot is moved horizontally and vertically. This allows all of the picture to be sampled as a set of lines. Most TV systems use interlaced scanning, meaning that all odd numbered lines of the picture are scanned, followed by all even-numbered lines*

scanning system, the even-numbered and odd-numbered lines of a picture are scanned alternately. Taking the UK system as an example, each scan takes 1/50 second so a complete picture (odd and even lines) is scanned at the rate of 25 pictures per second, but the half-scanned pictures repeat at 50 times a second. This greatly reduces flickering while keeping the bandwidth down.

Though early attempts at television used mechanical systems of scanning, closely following Nipkov's work, Campbell-Swinton in 1908 realized such efforts were a dead-end, and his outline of a purely electrical system was the path taken by teams in the USA and the UK in the years 1930 to 1937. The UK team at EMI was led by Isaac Schoenberg; that in the USA by Zworykin and, notably, both pioneers had studied with Boris Rosing, who had developed Nipkov's principles. The UK team determined that the image should be scanned using 405 lines, with a complete scan 25 times per second; the US team opted for 525 lines and 30 scans per second. In 1968, the UK adopted the newer European standard of 625 lines, 25 frames per second, which is still in use. Note that differences between scanning rates and in colour systems make it impossible to interchange video cassettes between countries with different systems.

The scanning on modern systems is electronic, using a beam of electrons in the receiver and in some cameras. Miniature cameras use a tiny array of light-sensitive cells (termed CCD, charge-coupled devices) scanned by sampling their outputs using techniques originally developed for computing purposes. Miniaturization of equipment and replacement of older transistor circuits by systems based on integrated circuits (silicon chips) has had the dual effect of making systems very much more reliable though almost impossible to service except by experienced technicians with access to a complete set of spare parts. The days when a piece of video equipment could be put back into service after some work with the soldering iron are more-or-less over, and though some experience of amateur electronics is always a useful attribute for the AV technician it is no longer such an important requirement as it once was.

Video signals, comprising of the three components of audio, brightness (luminance) and colour (hue) require a frequency range enormous in comparison with the 100 hertz to 10 kilohertz required for audio alone. The usual requirement for video is the range from zero frequency to 5.5 million hertz (5.5 megahertz), some five hundred times the range needed for audio.

It is this requirement which makes recording of video signals such a technical achievement. Audio tape recorder design shows increased frequency range can be achieved only by increasing tape speed past the head or by decreasing the gap size in the head. With a gap now at a size which is imperceptibly small and impossible to reduce further, the only way of increasing the recorded frequency range is by increasing tape speed. In the early days of video recording, the method used to give this increase in tape speed merely pulled much more tape passed the head at high speed, devoting several miles of tape to a 30 minute recording. Later technology adopts two other options.

The most recent option, termed *S-DAT* or *DASH*, uses multiple heads, making use of miniaturization techniques learned from IC technology. Some 12 or 28 heads are used with tape 0.25″ or 0.5″ wide (12 tracks on the narrower tape, 28 on the wider tape). The signal to be recorded is then split up among the different tracks, so tape speed is held down to something of the order of 75 cm per second. This should be compared with the standard 4.76 cm per second of audio cassettes. This type of equipment is, at the time of writing very rarely seen in cassette form and is not likely to be used in educational, general business or domestic video work – its main advantage is easier tape editing.

The more traditional option, if a technology only a few years old can be described as traditional, uses rotary heads. Instead of moving tape past a head at high speed, the head is moved past the tape. Further, by making the head movement at an angle of about 5° to the tape, the whole width of tape can be used to accommodate recorded tracks so that the total length of track is much greater than length of tape, as shown in Figure 5.2. Use of a 5° angle means each track across the tape is about 100 mm long – the

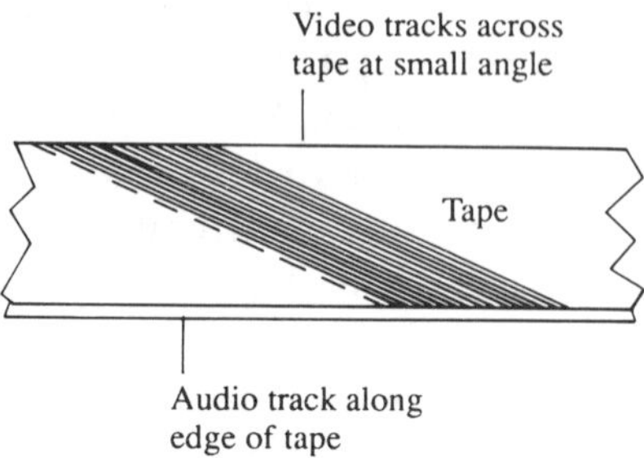

Figure 5.2 *Video recorders use revolving heads which record tracks across the tape at a shallow angle rather than along the tape. This allows fast movement between head and tape combined with relatively slow tape speed of a few centimetres a second*

illustration shows a rather greater angle in order to emphasize the slope of the tracks. This system uses two or more recording/replay heads located on a revolving drum. Figure 5.3 shows a simplified version of a standard VHS arrangement in which tape is

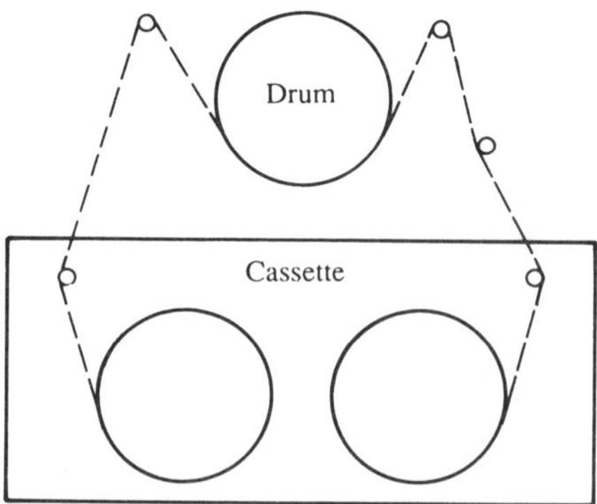

Figure 5.3 *A simplified view of the tape path in a video recorder. Some early VHS machines use paths almost of this pattern*

pulled out of the cassette and round a drum tilted at an angle of 5°, with two recording heads mounted on the rim of the drum. One head traces a recorded track on the tape and, as it leaves the tape, the other head starts to trace its track with a small overlap, ensuring one head or the other is always tracing a track. Video signal is split into separately-recorded parts, with the luminance (black and white) as one portion and chrominance (colour) as the other. Conventionally, audio signals are recorded along the edge of the tape, using the part of the tape where the signal recorded by one video head is identical to that traced on the other edge by the other head. In other words, one edge contains redundant video signals which can be replaced by conventionally recorded signals. This method is now fast disappearing in favour of using the revolving video heads for audio recording too. Much higher quality sound can be recorded in this way.

Mechanically, all this is done using an arrangement such as that shown in Figure 5.4. The illustration is of a Sony *Video–8* machine,

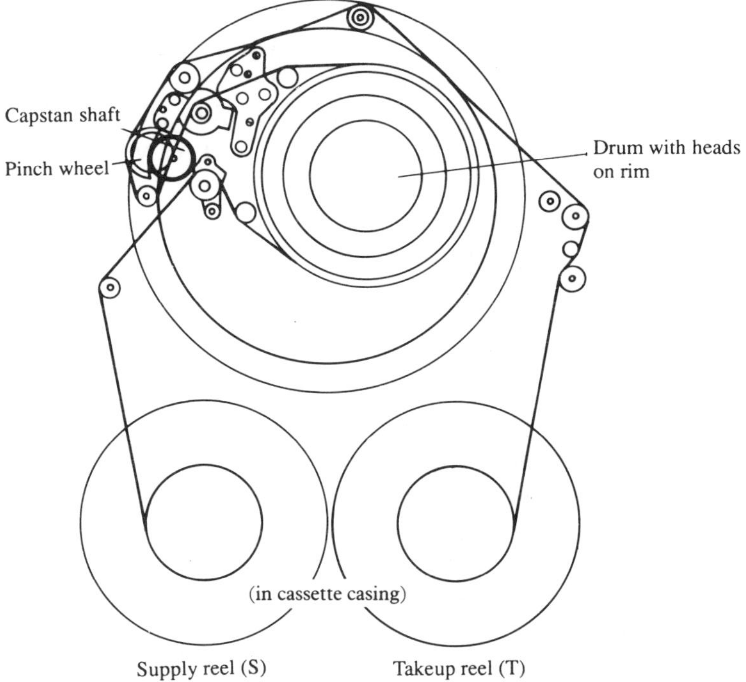

Figure 5.4 *The tapehead mechanism of a Video-8 recorder, showing the rotating drum which carries the heads, and the path of the tape which has been pulled out of the cassette. The movement of tape is controlled by a conventional capstan and pinch-wheel, and the audio track lies along the edge of the tape, erasing redundant parts of video tracks in this section (courtesy Sony UK)*

but principles are the same as those of earlier types, notably *Betamax* recorders. Tape is pulled out from the cassette over a drum angled slightly with respect to the tape to provide the angle for scanning the heads over the tape. Heads are built into the drum, which revolves at a steady high speed (typically 1500 revolutions per minute, corresponding to the TV frame repetition rate of 25 frames per second). Video signal is combined into one signal along with two audio signals (for stereo sound) and recorded in one track. This track also contains signals which synchronize the main video signal and which ensures speed of head rotation when the tape is replayed matches speed during recording.

Tape lacing is completely automatic and on *VHS* recorders is done each time the play or record control is used. Tape is unlaced for fast winding, except for the lower-speed form of fast-forward or reverse which allows pictures to be viewed. *Betamax* recorders in contrast, keep tape laced at all times until the cassette is removed. There are several variations on the *VHS* technology (which almost completely dominates home recording in the UK) including the *VHS-C* used for some video cameras, but the different versions retain an acceptable amount of compatibility. The *VHS-S* type permits a much higher picture and sound quality to be achieved while retaining the capability of playing tapes recorded on earlier machines.

Two other formats found are the older *U-Matic* and new *8 mm* varieties. When video recording first became available for other than studio use, the Sony *U-matic* system, using open reel wide tape, was the dominant technology, and is still in use today. This very enduring system makes recordings of markedly better quality than can be achieved with any cassette system, and is particularly useful if many copies of a recording have to be made. It permits editing to be done, and is in every way geared to the requirements of a small TV studio. *U-Matic* machines were the initial choice for AV centres, with recorded signals sent by cable to remote monitors, and with reasonable maintenance working life of such machines can be very long.

By copyright agreements, educational establishments are allowed to record broadcast TV programs and to keep recordings for 30 days, after which they must be wiped. Only a few centres which teach television techniques need to make original tapes which are then duplicated so, for everyday use of video recording, video cassette recorders are completely adequate – with the advantage of using standard technology, which can be serviced almost everywhere and uses widely available cassettes.

This standardization does not extend to video cameras. Unlike recorders or monitors, a camera generates all scanning and synchronizing signals to accompany the picture signal. It also generates the video signal from whatever image it is pointed at, preferably focusing and adjusting lens aperture automatically, and it records the combined video and synchronizing signals by some method allowing easy playback.

These requirements have been met in a wide variety of ways. One approach is to design the camera around a standard sized *VHS* cassette, so the camera is virtually an extension of a *VHS* recorder. This allows cassettes recorded by the camera to be replayed on any *VHS* recorder, but it forces the camera to be quite large and cumbersome. The second option is to use a miniature cassette (*VHS-C*) whose tape is recorded in *VHS* format and can be replayed by inserting it into a casing playable in a standard *VHS* machine. This allows camera size to be greatly reduced while still allowing tapes (of up to 30 minutes duration) to be replayed conventionally. This is the approach used, for example, in the Amstrad video camera whose price is so attractive but which is restricted in use to domestic purposes by its fixed-focus lens with no built-in monitor facilities.

The third option is to use a non-compatible tape format, and to design the camera so it acts as a replay device. This allows use of miniature cassettes using formats incompatible with full-sized *VHS*, but it ties up the camera when replaying is needed. This is of little importance when the camera is used at fairly infrequent intervals, and cameras of this type are now more common.

Currently, the most likely approach for the future, already available, is 8 *mm* equipment. Developments in tape technology, aimed both at video and at digital sound recording (the tape equivalent of CD, which allows both recording and replaying) have resulted in 8 *mm* wide tapes (compared with the 12.5 mm of *VHS*) in miniature cassettes. These can be used in cameras of quite incredibly small size, comparable in size with 8 *mm* film cameras, and for which video recorder/players are currently available. Equipment is much more familiar in Japan, USA and in Europe at present, mainly because UK users plumped so firmly for *VHS* it will take some time to change to the later 8 *mm* standard.

One particular advantage of 8 *mm* camera equipment is reduced battery drain. Older cameras required a separate battery pack which was large, heavy and expensive, and even the later smaller *VHS-C* types require a fairly large battery pack. The usual type of battery used is the nickel-cadmium type, and these present problems familiar to many users of portable equipment. Nickel-cadmium cells thrive on being used to their limits. Ideally, they should be drained in a session, recharged, and then shortly afterwards drained again. Unfortunately in many applications, cells are only partly drained, and are then subjected to recharging over long periods. This light use may result in battery capacity being greatly reduced, so it is capable neither of sustaining long periods of use, nor of retaining charge for more than a few days.

Where the battery pack of a video camera is used intensively, its life is normally long, though it is quite common for one cell to degenerate suddenly. The type of use in which the camera is used for ten minutes each Wednesday afternoon is the most destructive for cells and when there is a pattern of such use, the best way of preserving battery life is to remove it after the camera has been used, discharge it fully using a load such as a car headlamp bulb, then re-charge it before the camera is due to be next used. This carries a risk that the battery is not ready for use if the camera is needed at short notice, and the only way around the problem is to require a 24-hour booking period for the camera.

User guide

Video equipment represents a larger investment than most audio equipment, so that it needs to be treated with care, particularly a video camera. Users need to be acquainted with monitor, recorder, camera, and cassette care. Familiarity with television receivers and domestic recorders should make use of AV video equipment comparatively easy, but a surprising number of television viewers put up with receivers which are grossly maladjusted and, given a correctly adjusted set, usually manage to lose correct settings within a day or two.

Very often, the AV technician ensures all adjustments available to the user are correctly set, but each user should know how to return to correct settings if these have been altered. It is not unknown for the first person into a classroom to alter all settings on any piece of AV equipment, in the hope of gaining by the subsequent confusion.

Monitors feature much the same adjustments as a TV receiver, and usual accessible controls are audio volume, brightness, contrast and colour. In addition there may be a vertical hold or frame-hold control, used only if the picture appears to be rolling vertically instead of locked steadily. Audio volume is used as it would be on an audio system, but the three controls causing the greatest number of problems are brightness, contrast and colour, because their actions are interdependent and they should, apart from making small adjustments, be altered in sequence.

Correct settings, if these have been completely altered, are obtained as follows:

- Turn down colour control setting to obtain a completely black and white picture.
- Turn down brightness and contrast controls. View a picture (from recorder or other source). Adjustment is much easier if a recording of a test signal can be used.
- Adjust brightness so the black areas of the picture are *just* black. This may leave bright areas looking dull.
- Adjust contrast until contrast between black and white is acceptable. Too great a contrast makes pictures of the 'soot and whitewash' variety, with no shades of grey visible.
- Re-adjust brightness to keep black areas looking black.
- Repeat steps 4 and 5 until best range of shading is achieved. This is notoriously difficult to achieve on a changing picture, so a recording of a test signal is recommended.
- Adjust colour control only when a good monochrome picture is obtained. Failure to turn off the colour when adjusting the other controls is the usual cause of difficulty in attaining correct settings.

The video recorder needs virtually no adjustment in normal service, and the only adjustment normally provided (mainly on older machines) is track alignment adjustment, typically a wheel control as shown in Figure 5.5. This has its normal central position marked, and it is important to ensure this setting is maintained, because it affects both record and replay of cassettes. The effect of this control is to alter slightly head position relative to tape, and it is intended to deal with the situation where a cassette is recorded on a machine whose heads are positioned differently. This control was put into recorders in the days when manufacturing tolerances were not so tight, and it is unusual to use it. The symptom of incorrect head positioning is the appearance of a jagged line across the screen, usually at the top or bottom of the screen. Something similar can also be caused by a monitor fault (line tearing), so if adjusting the track alignment control has no effect the technician should be informed. If the bar

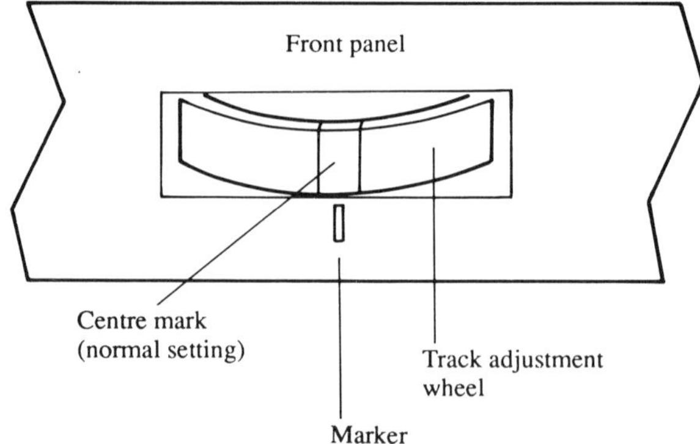

Figure 5.5 *A typical track alignment adjustment control as used on older recorders. This is intended to cope with variations in head alignment between different machines, but in practice is seldom needed*

is due to bad head alignment, control adjustment moves the bar so it can be moved off the screen.

In the unusual event of this being required, a note should be made of the cassette causing the problem, and the track alignment control should be returned to its normal position after the machine has been used. If the adjustment is maintained in its altered setting, all cassettes subsequently recorded on this machine require this setting to be used. The reason for noting the cassette is that this may have been recorded on another machine whose head alignment may have been adjusted in error (or in malice), or which needs servicing.

The camera

The following description applies to the Sony *CCD-V200E Video–8* camcorder, but is in general applicable to most types of modern video camera equipment. This uses two types of battery, a nickel-cadmium battery for main power and a lithium battery for operating displays and controls – this latter battery has a long life (about two years). Power loading on the nickel-cadmium battery is about 8.5 W, and the camera can also be operated, using adapters, from AC mains or from a car battery.

A modern video camera in this class positively bristles with sockets, switches, meters and adjusting buttons (66 in all on the

Sony camera) so it is too much to expect any user to become acquainted with the full gamut of controls in the course of forty minutes use on a Wednesday afternoon. That said, the actual number of *essential* controls, as distinct from available, is quite small, particularly if camera use is restricted to well-lit scenes of a straightforward nature, and the AV technician presets other controls. The viewfinder uses a miniature TV (black and white) screen rather than using the lens directly, so what is seen in the viewfinder corresponds closely with what is being recorded. Note some low-priced cameras use a simple lens viewfinder, with little correspondence between what is viewed and what is being recorded.

Three actions in particular are available in automatic form to make use of the camera simple. These are focus, white balance and iris. The autofocus uses an infra-red beam from the front of the camera and is reflected from the subject used, rather like a radar beam, to calculate distance and adjust focus accordingly. This means the camera is focused on the subject which occupies the central area in the picture and, though suitable for many types of shot, it can be unsuitable for others in which the main subject is to one side. Infra-red focusing systems can also be upset by the use of other infra-red devices such as TV, video recorder or slide projector remote controls. Some cameras use an autofocus beam projected and received through the main lens, so allowing auto-focus over a wider range of distances – particularly useful when zooming is used.

Auto-focus is upset by a fairly wide range of subject matter and for video cameras, unlike film cameras, these subjects can be dealt with only by switching to manual focusing. Typical problems are caused by:

- Black or very dark subjects.
- Slanting surfaces from which the reflected beam cannot return to the camera (room walls, floors or ceilings).
- Distant absorbing materials (foliage).
- Excessively reflecting materials (mirrors).
- Non-solid objects (smoke, fireworks).
- Very small subjects (insects).
- Subjects with two reflecting surfaces at different distances (animals in cages).
- Continually moving objects (runners, cars).
- Objects which emit infra-red (automatic doors, lights).
- Distant objects (more than 10 m away).

Always switch to manual focusing if in doubt.

Colour balance control is intended to ensure correct colours are shown irrespective of whether the illumination is sunlight or filament light. As light from filament lamps is yellow, indoor scenes would show a predominantly yellow cast if the balance control is not altered from the outdoor setting. Cameras intended for amateur use do not permit a complete range of colour balance control as this requires instruments to set up, and the usual choice is indoors, outdoors or automatic. One point to note is the sensor for the light is at the camera, so if the camera is set to AUTO colour balance and held indoors with artificial lighting but filming an outdoor scene, colour balance is incorrect. For such a case, outdoor setting rather than auto setting is appropriate. Conversely, if the camera is being used outdoors but filming through a window into a room with artificial lighting, the indoor setting should be used.

Iris control determines the amount of light entering the lens and affecting the pickup-device. The automatic setting becomes unsuitable if there are conditions of extreme contrast, say, a bright light is behind the subject making the subject look dark, or because the subject is strongly lit making the rest of the scene dark. For a subject too dark because of back-lighting (shooting into the light), the iris should be manually opened so detail of the subject can be seen. For a brightly lit subject in a pool of darkness, the iris should be manually closed. The control can be returned to automatic after such scenes have been shot.

A control more specialized but which needs to be mentioned is shutter speed. Normally, shooting speed of the video camera is fixed at 25 frames per second, as for normal television, but many cameras allow variation. A slower shutter speed produces a tape which when played at normal rate, presents a speeded-up appearance; conversely using a fast shutter speed gives a slowing-down effect when the tape is played. This allows the camera to be used for purposes such as slow-motion analysis of athletes and gymnasts or speeded up shots of insect-eating plants or racing tortoises.

The sequence of use is fairly simple if automatic settings are for all controls. Assuming that batteries are inserted and a cassette is ready for action, with cassette speed switched to normal (rather than long-play), letters refer to the illustrations of Figure 5.6:

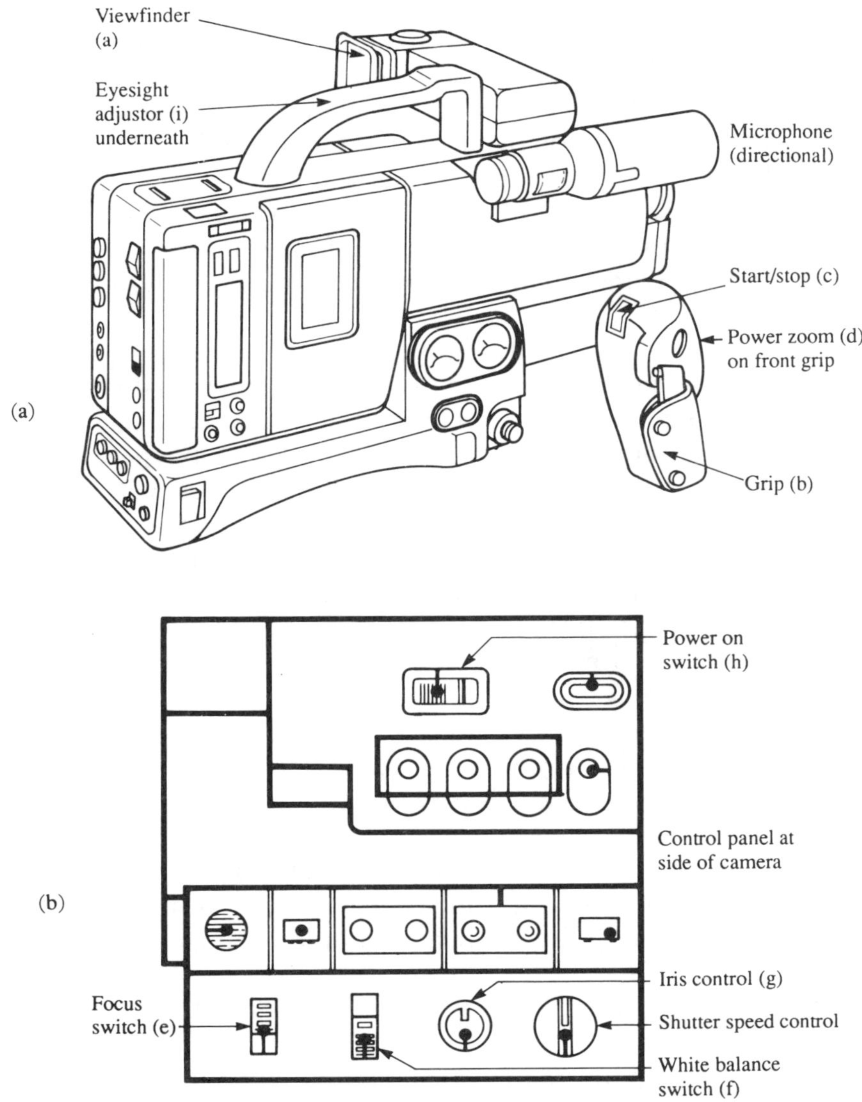

Figure 5.6 *Some of the controls on a camcorder. A Video-8 model is illustrated, but the main controls of most cameras follow this pattern (a) main body (b) other controls on a panel at one side of the camera*

- Fold out the viewfinder (a) and the grip (b).
- Hold the grip so thumb and forefinger can operate the Start/Stop button (c) and the power-zoom button (d).
- Set focus (e), white balance (f) and iris (g) selectors to the automatic positions (green on the Sony *CCD-V2000E Video–8*).

- Switch camera power on (h) and remove the lenscap.
- Adjust viewfinder lens (i) to suit your own eyesight.
- Point to the scene, use the zoom (d) to obtain whatever magnification you need, and press (then release) the start/stop button (c) to record the scene.
- To pause, press the start/stop button again. Some cameras require this button to be pressed for recording.
- Turn power switch off when recording is complete.

This very simple outline covers a surprisingly large percentage of video camera uses, but the essence of a camera in this class is that it can cope with requirements well outside these uses. Such refinements are outside the scope of this book; they would provide material for a manual of their own, and most camera handbooks provide the information, though often in compressed form. For anyone intending to make serious use of the video camera, a good background in amateur movie-making is very desirable, and some of the books published in the heyday of Super–8 movies contain advice which from both technical and artistic points of view is excellent.

As most users of the camera are likely to use it in the fully automatic mode, a few hints relating to the *Video–8* types, but applicable to most, can be very helpful:

- Be careful not to cover auto-focus apertures – there are two of these (one beam-out, one beam-in) at the front of the camera. This is not applicable if the main lens is used for auto-focus beams.
- Do not encourage any subject being filmed to look at the auto-focus beam.
- Do not grip the manual focus control while auto-focus is in use.
- Sudden changes in light intensity can cause odd effects in the viewfinder.
- Auto-focus and white-balance do not operate correctly in poor light.

Technical maintenance and operation

Maintenance of video equipment consists of cleaning, maintaining correct adjustment, minor repairs, care of cassettes and battery charging for cameras. In addition, the technician needs to be able to clean the heads of video recorders, the most commonly needed maintenance on machines which are intensively used. The technician is also responsible for making video recordings from broadcasts, keeping a log of such recordings and ensuring programmes are wiped from cassettes in the specified time.

The main day-to-day activity is the making of recordings and care of cassettes. Each cassette should carry a label showing date when recorded, subject, and date shown – the computer can be pressed into service to print suitable labels. Cassettes should be stored so new cassettes and old recorded cassettes ready for recycling are kept in one place, cassettes recorded but not yet used in another, and cassettes recorded and used and not yet ready for re-use in a third position. Another useful system is to stick a red dot onto cassettes which have been recorded but not yet played, and to remove the dot whenever cassettes have been used.

A booking system for making recordings is essential. In the early days of making educational video recordings, when recorders were not fitted with time switches, it was necessary for the technician to remain with the recorder and this could be troublesome when an evening recording was wanted. Modern video recorders have solved this problem, but the uncertainty of scheduling which grips broadcasters when a sports competition looks like taking up more than 23 hours' television in a day can occasionally cause timing to be unreliable, and though it is best to set starting time for a recording to the scheduled time, the ending time should make a generous allowance for the previous programme running over.

If there are cassettes of material which must be kept, the record-protection tab on the cassette should be broken off and they should be labelled in a different way, storing them separately from the general run of cassettes, which will be re-cycled.

The main maintenance operation of video recorders is head cleaning, and as in the case of audio recorders, this should be done using felt pads moistened with isopropanol rather than by using cleaning tapes. The frequency of head cleaning depends on the extent to which machines are used and varies considerably. The method of cleaning is straightforward, but gaining access to the heads is less straightforward on some types of recorders. In outline, the method is as follows, using a chamois-leather pad supplied by the maker of the equipment:

- Remove the cassette, and disconnect the recorder from the mains.
- Remove the screws holding the top cover of the machine in place, and lift up the cover. On some top-loading machines, it may be necessary to open the loading slot before disconnecting, and it may require some manipulation of the top cover to remove from around the loading slot. Front-loader machines are easier to work on.
- Locate the head drum (Figure 5.7) and find the head positions. Make sure there is no possibility of heads coming into contact with any hard materials.

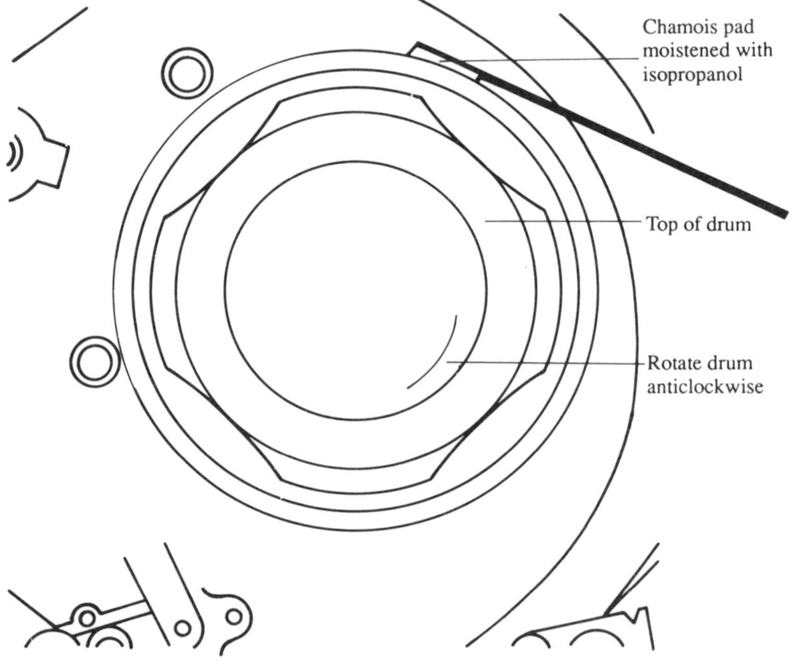

Figure 5.7   *The head-drum of a recorder or camera being cleaned, using a chamois-leather pad moistened with isopropanol. Note the pad **must not be rubbed** up and down the head drum, only held in place while the drum is revolved anticlockwise. The reel tables (on which the cassette rests) must be cleaned, and drive belts should also be cleaned to remove any trace of grease*

- Moisten the chamois-leather pad with isopropanol and hold it against the drum surface. Slowly rotate the drum assembly anticlockwise with one finger of the other hand. *Do not turn power on, or rotate clockwise and do not rub the leather against the drum.* Degauss the heads if the maintenance schedule calls for this to be done and if a suitable degausser is available.

- With the cassette compartment in its ejected state, clean tape guides (all of them – 11 on the Sony *8 mm* recorders), capstan spindle and the pinch wheel (Figure 5.8) using chamois-leather moistened with isopropanol.
- Clean the timing belt and reel-table surfaces (Figure 5.9) with a cloth moistened with isopropanol.
- Allow time for all traces of isopropanol to evaporate before assembling the top cover again.
- Connect the machine up and test it.

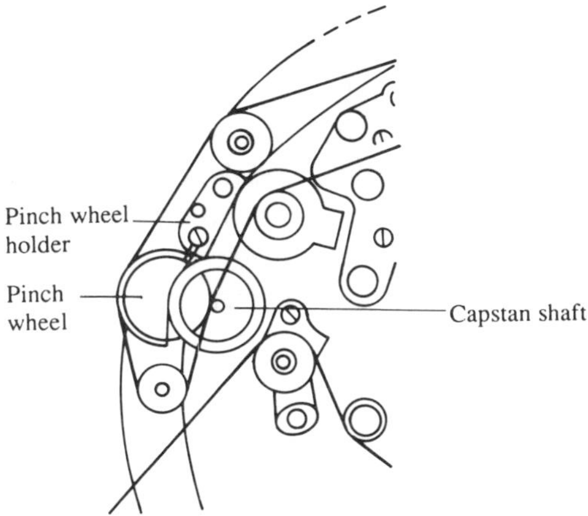

Figure 5.8   *The tape drive uses the familiar pinchwheel and capstan construction to ensure even tape speed. These parts require cleaning at intervals*

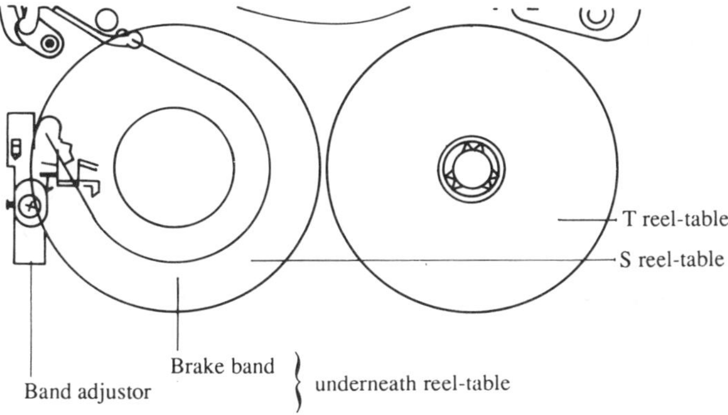

Figure 5.9   *The reel-tables on which the cassette reels rest must be checked at intervals*

While the top cover is removed, check that all fastenings are tight and ensure that the whole area around the head drum is clean and free from tape dust. Any cleaning of dust should be gentle – a miniature vacuum cleaner (with a soft plastic nozzle) is ideal, but a soft brush is also acceptable. Dust removal should preferably be done before head cleaning so as to avoid having dust adhering to wet patches of isopropanol. Lubrication should be carried out only in accordance with the makers instructions and using the recommended lubricant.

Head life is often quoted as 2000 hours, corresponding to about three years use at four hours per day, five days per week in school and college term times. In fact, recorders are rarely used quite so intensively and head lives seem to be in excess of the quoted 2000 hours, anyway, so head replacement is not a common requirement. If the quality of picture deteriorates and is not restored by cleaning the heads, however, either head replacement or re-alignment of the head or the guide pins may be necessary. These are specialized jobs; though head replacement on modern machines is not difficult it has to be carried out in conditions of scrupulous cleanliness.

Camera maintenance

The main maintenance action on cameras is battery charging, and the remarks on use of nickel-cadmium batteries earlier should be noted. Never leave batteries to charge over a weekend – quite apart from the fire risks of unattended equipment, this may over-charge batteries and make them less efficient. Unlike lead-acid batteries, nickel-cadmium types come to no harm by remaining discharged for long periods, and are best kept in the discharged condition if they are not to be used for some time. Batteries should be discharged before holidays, for example, using a load such as a car headlamp bulb (some load capable of dissipating the power at the battery voltage) – there will be no harm in dissipating a 9 V battery pack into a 12 V bulb, but if you need to discharge a 24 V pack, use two headlamp bulbs wired in series.

Charging of nickel-cadmium battery packs is usually taken care of by chargers supplied with cameras. Watch, in particular, for a bad cell in a pack. This is detected by measuring the voltage of each cell while the cells are on load (passing 0.5 to 1.0 A of current), and looking for any one cell whose voltage is notably lower than the others. A bad cell can sometimes be revived by the circuit shown in Figure 5.10, charging up the electrolytic capacitor then discharging it through the cell suddenly. This should be done when the cell is

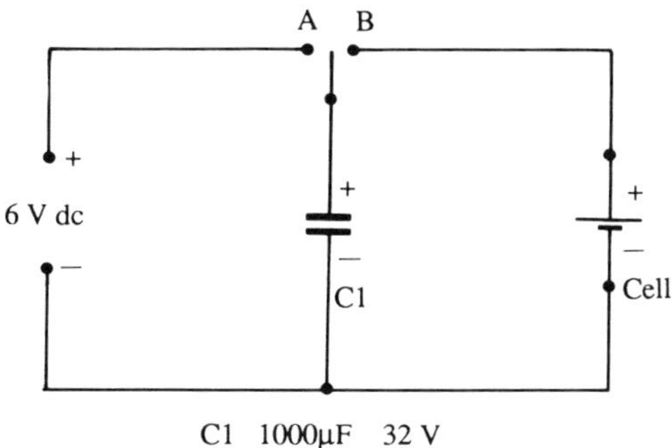

Figure 5.10   *A circuit which can be used to resuscitate a failed nickel-cadmium cell*

completely discharged, and repeated a few times. The cell can then be put into a single-cell charger unit to see if it then recharges correctly. This is a very useful procedure which can triple the life of a cell, and it can be a considerable saving when cells in a battery start to fail one by one.

Other camera maintenance procedures concern lenses and cassettes. Lenses should be cleaned with optical tissue after removing any loose dust or dirt with a soft brush. Photographic dealers sell suitable brushes attached to rubber bulbs, allowing a puff of air to be directed at the dust along with the brushing. This obviates the problem sometimes arising where dust is moved around the lens but not so easily removed. Never try to dismantle lenses, which consist of several optical components along with the mechanical iris system for controlling light aperture. Follow any special cleaning and lubrication requirements for zoom lenses.

Camera recording heads need to be cleaned just as heads of recorders are cleaned, but in educational or business applications it is unlikely a camera receives such intensive use as a video recorder, so head cleaning need not be so frequent – it may not even be necessary in the working lifetime of the camera.

Hints and tips – mainly camcorder

- Never use any video equipment (recorder or camera) in conditions of high humidity, particularly when equipment is cold. Most recorders and cameras can sense condensation and will refuse to operate in such conditions because of the risk of tape sticking to the rotating drum.
- On a camcorder, always run a new tape for 15 seconds in playback mode before using for recording. This ensures that you start on a piece of tape which is not right at the start of the reel.
- Never insert anything into the small holes at the rear of a *Video–8* cassette, because these holes are used for sensing tape type.
- Be careful of lithium batteries – return to the manufacturer for disposal, since lithium is a fire hazard if it comes into contact with water entering a broken battery-casing.
- Some cameras allow use of both a subject and narration microphone. Many permit limited editing facilities on tape within the camera.
- Use a wind-noise filter switch, if provided, when recording outdoors in other than calm conditions.
- In bright light conditions, it is often possible to operate with fixed values of zoom and focus, avoiding problems of autofocus.
- When operating outdoors at night under sodium lights (yellow light), set colour balance to indoor position.
- Zoom microphones are obtainable – useful for wild-life sound recording, in particular.
- Many cameras switch off automatically after a long pause.
- The more the camera moves, zooms and pans, the more amateurish your efforts will look. Do not treat the camera as a hosepipe.
- If a head does not respond to cleaning with chamois-leather and isopropanol it can, as a last resort, be scraped clean with a plastic spatula or the edge of a visiting-card.
- Always keep lenses capped when not filming.

6

Public address and sound reinforcement

Public address systems are not generally thought of as part of the AV range, but the need to be heard by a large number of people often arises in an educational or business context, many schools and colleges make provision for loudspeakers to carry emergency messages, and in conference centres the PA system is an important part of the AV stock in trade.

In addition, where education includes performance of plays and music from a stage before a large audience, some form of sound reinforcement may be necessary, particularly if the hall is not well designed from the acoustic point of view (and very few are). The use of the bull-horn type of megaphone for sports coaches might also come under the wing of the AV department, and so also would various intercoms for use between widely separated offices. This is the scope of devices covered in this chapter, but the emphasis is on use of microphones and loudspeakers in halls and large rooms, as this is also the area that presents the greatest difficulties.

Principles

Principles of public address and sound reinforcement are far from simple – as can be observed at railway stations and airport lounges. It is one thing to provide amplification of sound, but preserving intelligibility is quite another thing and something often overlooked when the AV department is asked to 'knock up something for Parents' night'. Designing a successful PA system is as much an art as a science, and demands considerable experience as well as appreciation of the factors of engineering, acoustics and human physiology that are involved. The first of these factors concerns the sound wave itself.

A sound wave is generated in air whenever any vibrating material is in contact with the air, and sound wave frequency is the same as vibration frequency, measured in units of hertz (Hz), or complete cycles of vibration per second. Sound is audible when the frequency range is within the capabilities of the ear (about 20 Hz to 20 kHz for a young human ear) and when sound amplitude, corresponding to the amount of vibration, is sufficient to operate the ear. Sensitivity of the ear is quite remarkable, corresponding to the ability to detect pressure changes of about ten millionths of a Pascal; 0.075 millionths of a millimetre of mercury. There is, thinking about modern music, some irony about measuring sound intensity in terms of a heavy metal.

The frequency range that the ear can detect reaches a peak in early adolescence and then steadily declines, so in the mid-fifties the upper frequency limit is about 13 kHz, and the decline persists throughout life even if no hearing problems develop. Elderly listeners are likely to appreciate more treble boost in replayed sound to compensate for this loss. The lower limit of hearing does not seem to vary much with age.

The response of the human ear is not uniform over the frequency range, and reaches a peak for frequencies around 3.5 kHz. This is the predominant frequency in female voices (which may be a pointer to something or other) and in practical terms means that intelligibility suffers if frequencies around this range are not prominent in sound. Figure 6.1 shows how the sensitivity of the ear for different frequencies alters for sounds of different amplitudes. This set of curves (derived from the well-known Fletcher-Munsen curves) shows that variations in sensitivity are least for loud sound, most for weak sounds. This is why a recording of music can seem so feeble when replayed softly, and points to the ideal for sound reproduction where such music should be replayed at the original level of the performance. Quite where this leaves instruments that create no sound (only electrical waves) is difficult to say.

The importance of all this for PA is if a high sound intensity can be achieved everywhere in a room, the sound can be of a broad frequency range, but if there is likely to be a large variation in intensity, with points where intensity is low, the range around 3.5 kHz should be emphasized.

The main problem affecting PA is reverberation. In an open space, most of the sound reaching the ear of a listener comes from the nearest loudspeaker. In a closed space, a significant amount of received sound comes as echoes, and this drastically affects

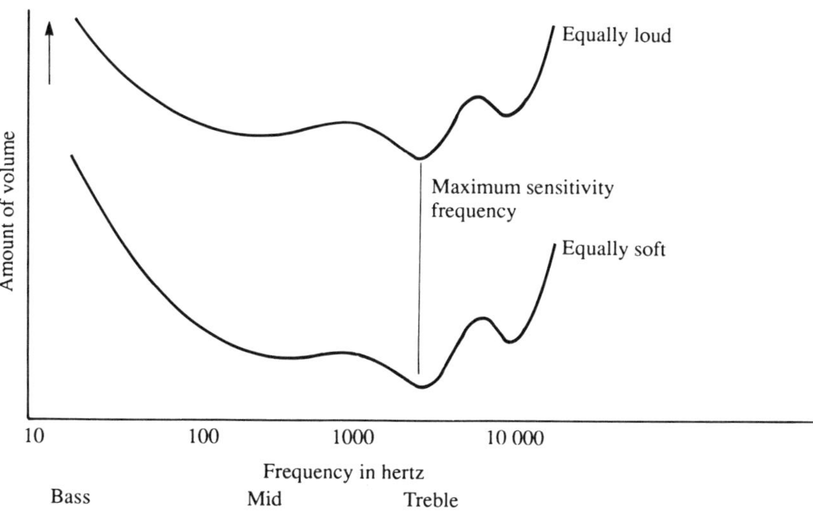

Figure 6.1 *Sensitivity of the ear at different levels of sound. The ear is most sensitive to sounds at around 3.5 kHz, very much less sensitive to low frequencies, and the pattern of sensitivity is different at different levels of sound at the ear*

intelligibility of sound, as anyone who frequents Cathedrals must have noticed.

The difference between a closed space and an open space has two effects. Firstly considerably more power is needed for public address in open spaces, because of the absence of echos. This requirement can be modified to some extent by using fairly directional loudspeakers, so there is less power lost in addressing the local sparrows. Nevertheless, large crowds outdoors require a large number of loudspeakers, and this in itself brings a problem of intelligibility akin to the echo problem.

Figure 6.2 shows the nature of the problem, with loudspeakers

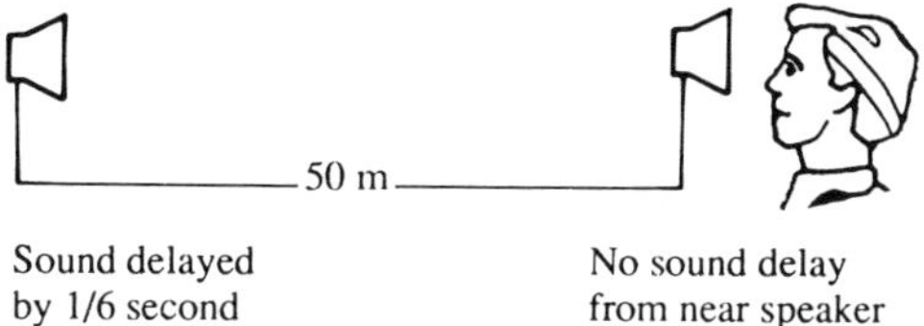

Figure 6.2 *When sound reaches the ear from two sources of identical sound spaced far apart, the delay in hearing the farther sound can cause unintelligibility*

placed 50 m apart in line. A listener close to one loudspeaker hears sound directly from it, and also from the other loudspeaker 50 m away. The sound from the distant loudspeaker, however, is delayed by the time it takes for sound to travel 50 m, about 1/6 second, so there is a blurring of the sound that reaches the listener. This blurring becomes worse if loudspeakers are more widely separated or if sound from, say, a third loudspeaker at 100 m distant also reaches the listener's ears.

Professional PA systems overcome this by deliberately delaying signals to loudspeakers, using the system shown in Figure 6.3. The loudspeakers are directional, so the predominant direction of sound waves is from A to B. The loudspeaker at B has its signals delayed, so a listener at B hears the two sets of sound in step while listeners at A hear only the sound from the loudspeaker at A because of the directional nature of the loudspeakers. The delay is achieved using devices called acoustic delay lines which can be obtained at quite reasonable prices, or by using tape loops for longer delays.

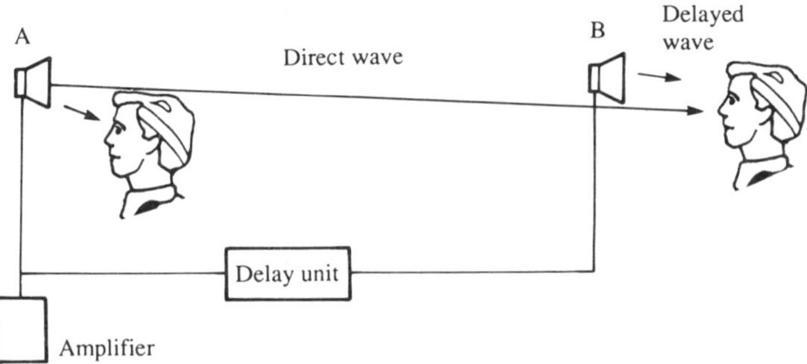

Figure 6.3   *The delay system that is used in PA systems along with directional loudspeakers to reduce the effect of different sound paths*

The more obvious solution, to use only one loudspeaker or set of loudspeakers at one sound source, is practicable only for small numbers of listeners closely gathered together. In a large gathering, setting the volume level so listeners at the back can hear results in deafening the listeners at the front, and also results in microphone feedback (see later, this chapter).

Sound systems in closed spaces are a different proposition, due to the effect of echoes. The worst situations are those in which there is a combination of free-space conditions, with listeners spread out over large distances, and echoes from a high roof and other buildings; the classic railway-station problem. Cloistered rectangles present similar problems. Fortunately, such situations are rarely encountered by AV departments so this removes the worst type of problems from our consideration.

Echoes in a hall or large room are created by reflections of sound waves from walls, ceiling and floor, and the extent of echoing depends very closely on dimensions of length, width and height. If we express both length and width as a ratio to height, it is possible to plot a diagram which shows the room dimension ratios which provide acceptable sound. Such a diagram, of fundamental importance in the design of PA systems, is illustrated in Figure 6.4, with the inner unshaded area indicating acceptable measurements. The centre of this area corresponds to a room with length about double the height and width about 1.5 times the height, so a room with a 15′ ceiling height is completely acceptable if its length is 30′ and its width is around 22.5′. At it happens, many halls conform to this set of ratios quite closely.

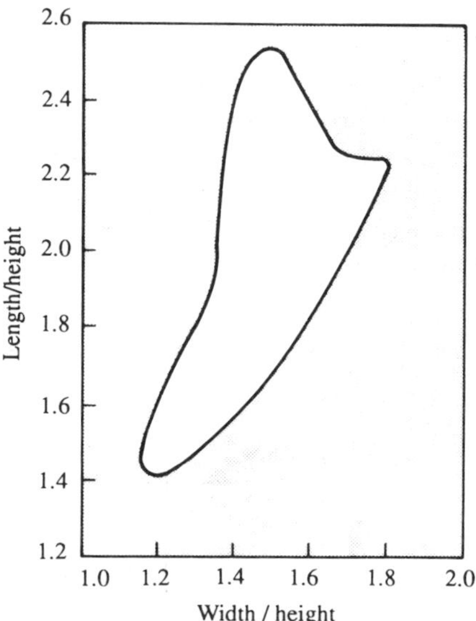

Figure 6.4   *Room dimensions and echoes. The ratios of length/height and width/height plotted here indicate a region, shown unshaded, of acceptable sizes*

The significance of the dimension ratio is that if ratios fall into the unacceptable region, artificial methods of controlling echoes need to be used if a PA system is to be intelligible, and even for unreinforced sound, voice or music, the hall is profoundly unsatisfactory for listeners. Echo control is not simple, and though some improvements can be made on a cut-and-try basis, professional design is needed to achieve really good results. Fortunately, it is seldom necessary to approach perfection in PA used in most contexts, only to mitigate the worst effects of an unsatisfactory building.

Reverberation time for a room or hall is the measurement which determines how effectively echoes are controlled. Accurate measurements of reverberation time are a matter for experts, but a rough idea can be obtained from listening to short percussive noises (sticks clattered together, for example) and estimating how long the sound takes to die away. It is generally agreed a reverberation time of 1.5 seconds or just less is desirable. A very short reverberation time can make a room sound 'dead' and unresponsive, very difficult to speak in or play music in.

Control of reverberation time is done by using sound absorbing surfaces and cavities. Absorbing surfaces are used to control predominantly treble echoes, and cavities to control predominantly bass echoes. Typical absorbing surfaces are curtains, foam plastic sponge material, acoustic tiles, and fibreglass. Of these, curtains and foam plastic are more acceptable in most establishments, though the hazards of toxic fumes from foam plastics in the event of a fire may rule these out. Heavy curtains are acceptable on all counts, and can be obtained at short notice if there is a surplus of funds at the end of the financial year.

The cavity form of absorber takes the shape shown in Figure 6.5, using a cavity lined with absorbent material. A cavity is of

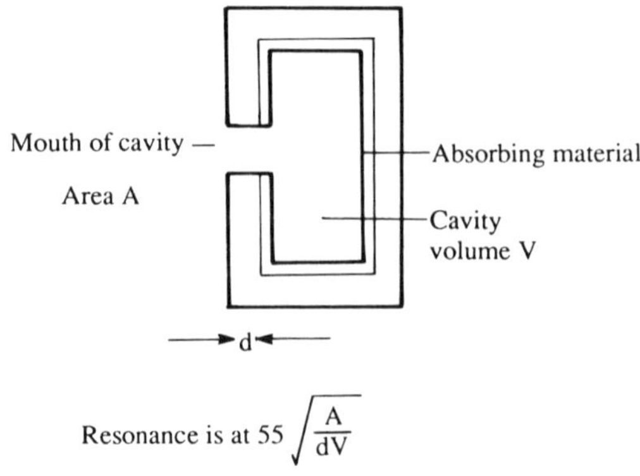

$$\text{Resonance is at } 55\sqrt{\frac{A}{dV}}$$

Figure 6.5 *A cavity form of absorber and its formula. Rooms which contain curtained alcoves often suffer from absorption where these alcoves act as cavity absorbers*

dimensions making it resonate to a bass note, and the absorbent material permits sound energy at that frequency to be absorbed (absorbed sound is converted to heat, but the resulting temperature rise is immeasurably small). The formula for cavity dimensions is shown – cavities to absorb low bass sounds are large and, in some halls, window recesses have a bass-note absorbing effect which is unwanted.

One of the main irritations of amateur use of PA is microphone feedback. Sound from the loudspeakers reaches the microphone to be amplified over again, and this effect, called positive feedback, results in the familiar howling sound whose pitch depends mainly on the distance between the microphone and the loudspeaker from which the main feedback is occurring. When this happens, the whole system acts as an oscillator.

A few commonsense precautions can cut down the extent of feedback, though its elimination requires more advanced measures:

- Use directional microphones and loudspeakers, pointing in opposite directions.
- Always place microphones 'upwind' of loudspeakers; so the order of placement is microphone(s), then loudspeaker(s), then audience.
- Avoid aiming loudspeakers or microphones at echoing surfaces.
- Keep the volume control of the amplifier system set low so microphone users need to be close to the microphone.

Sensitivity of microphones is particularly important where a roving microphone is being used to pick up responses from members of an audience. This microphone should be highly directional, shielded in foam, and used very close to the mouth of the user, otherwise feedback is almost certain to cause problems.

Where feedback is still a problem, in particular at sound levels used in pop concerts, electronic methods have to be used to eliminate feedback. These operate by comparing the fedback sound with the input to the amplifiers, and shifting signal phases (their relative timing) so feedback always acts in the reverse direction (Figure 6.6). This type of solution is not generally available to the AV user, though suitable circuits for construction have appeared in electronics magazines and can be constructed by an AV technician with an interest in and experience of electronics work.

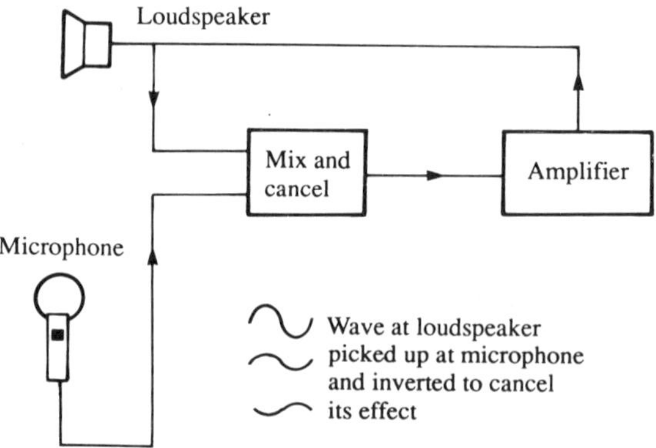

Figure 6.6 *The scheme used when microphone feedback cannot be avoided. The electronic unit compares sound at the loudspeakers with that at the microphone and continually alters the signal to the amplifier so the effect of sound picked up from the loudspeakers is cancelled out*

Loudspeaker types

The form of loudspeaker almost universally used in educational establishments and conference centres is the column, comprising a number of conventional loudspeaker units in a long slender enclosure as shown in Figure 6.7. Though this type looks pretty, its performance is usually poor, particularly where the individual drive units are of low quality. Where column loudspeaker units are installed, only a few listeners enjoy a satisfactory sound reception. The main reasons for their use are (1) they can be unobtrusive (2) they are easy to manufacture (3) they are comparatively cheap. It is possible to design effective loudspeakers of this type, but only on an individual basis, and certainly at a price much higher than most are prepared to pay.

Professional PA installations nowadays use a constant-directivity horn type of loudspeaker (Figure 6.8) which disperses sound over a wide angle with more uniform coverage than is possible using other systems. Such speakers also have a wide frequency range, typically 600 Hz to 16 kHz and are free from the effect, common with other types, in which higher frequencies are concentrated along a narrow axial beam. Units are large and do not conform to the box shape matching most halls, but their acoustic performance more than makes up for their looks.

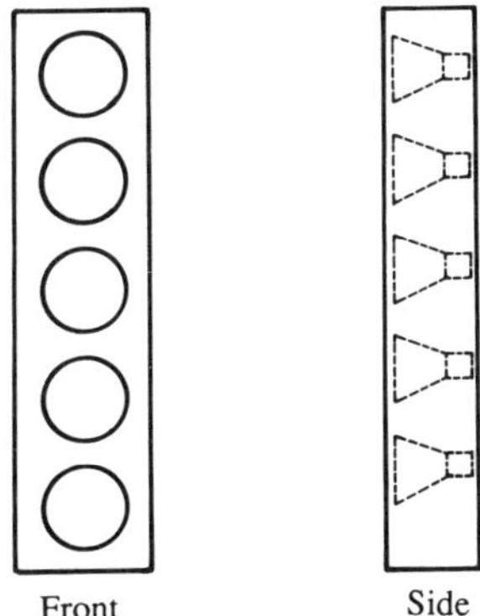

Figure 6.7 *A typical column loudspeaker array, consisting of a set of small conventional loudspeaker units in a long slim cabinet. This looks good but performs poorly*

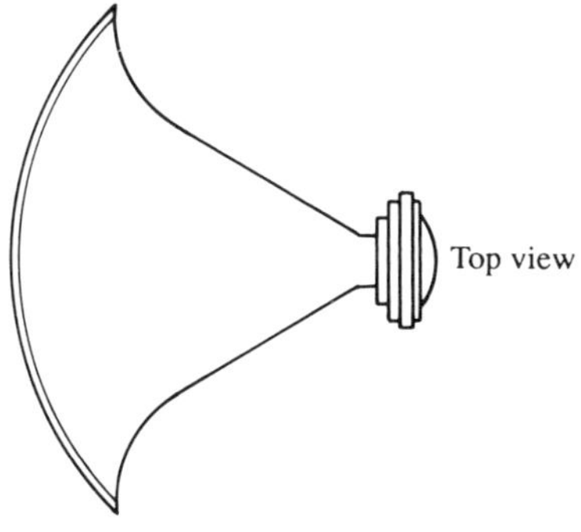

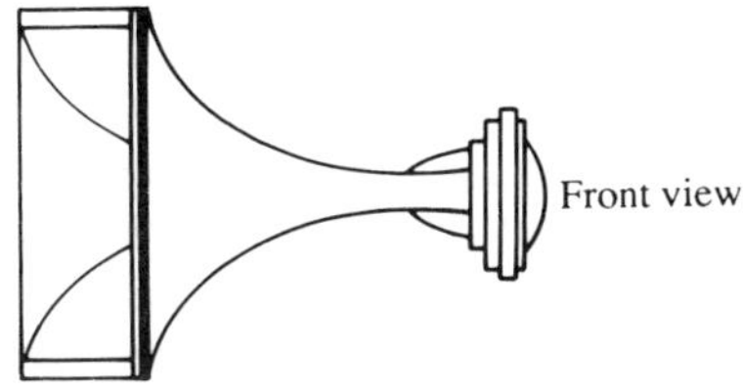

Figure 6.8 *The shape of the constant-directivity horn form of loudspeaker used in professional PA work*

Loudspeaker connections present a particular problem for makeshift arrangements. Like any other electrical devices, loudspeakers can be connected in parallel, in series, or in series-parallel

(Figure 6.9), and the arrangement used depends on number of loudspeakers, amplifier output impedance, and impedance of each loudspeaker. Typically a loudspeaker impedance for PA work is 8 Ω, and amplifier output impedance anything from 4 Ω to 8 Ω. Note that modern transistor amplifiers deliver more output power as loudspeaker impedance is decreased; there is no optimum impedance at which power is a maximum. If a large number of loudspeakers are connected in parallel, however, to produce a very low impedance, there is a danger that the amplifier will be overloaded.

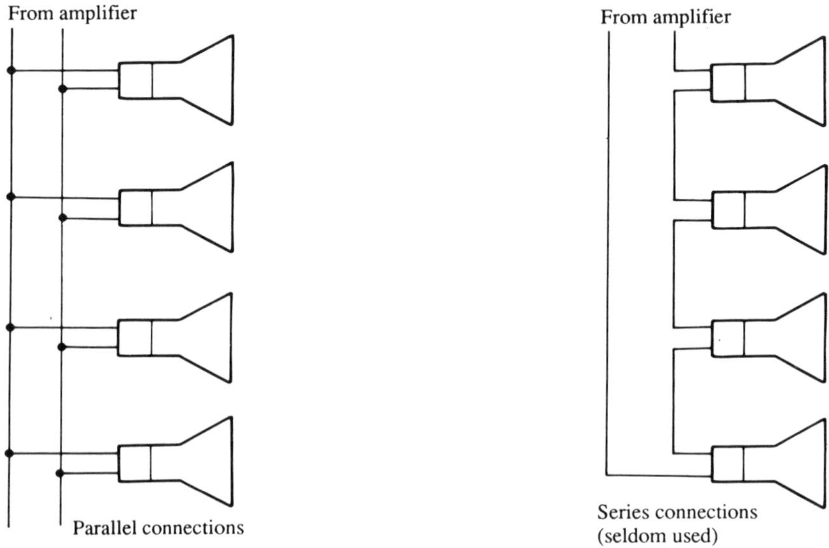

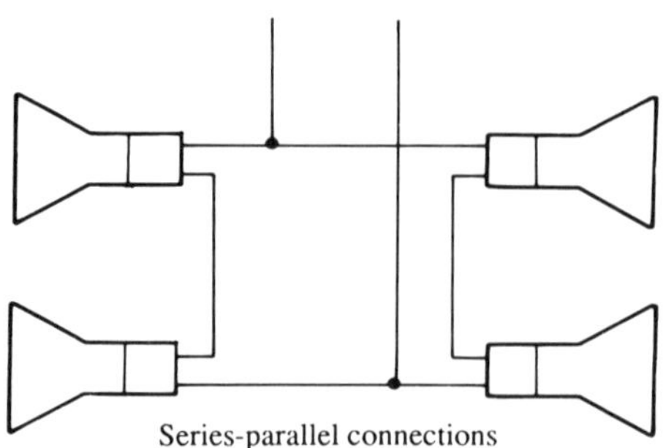

Figure 6.9   *Loudspeakers connected in parallel, in series and in series-parallel. Because total resistance of the loudspeakers must match that of the amplifier, series-parallel is the most common arrangement*

The sensible solution is to connect loudspeakers in such a way that the overall impedance is between 4 Ω and 8 Ω, and this is most likely to require a serial-parallel arrangement. Using loudspeakers of equal resistance (impedance) and power handling, numbers of loudspeakers can be connected so as to give the value of one single loudspeaker. This simple method makes use of numbers of loudspeakers which are perfect squares. If this cannot be done a configuration and its total resistance must be worked out from first principles – in addition there is a risk that some loudspeakers may take more than an equal share of the power. An option, used in professional PA equipment, is to use the 100 V line system (Figure 6.10). This requires an amplifier whose output stage uses a transformer which provides a signal at 100 V peak when the amplifier is operating at full rated power. The loudspeakers each use a step-down transformer, and are connected in parallel. The amplifier is fully loaded when the total power taken by loud-speakers equals the maximum output power of the amplifier – if the amplifier is rated at 100W it can support ten 10W loudspeakers if these can be connected to take equal amounts of power.

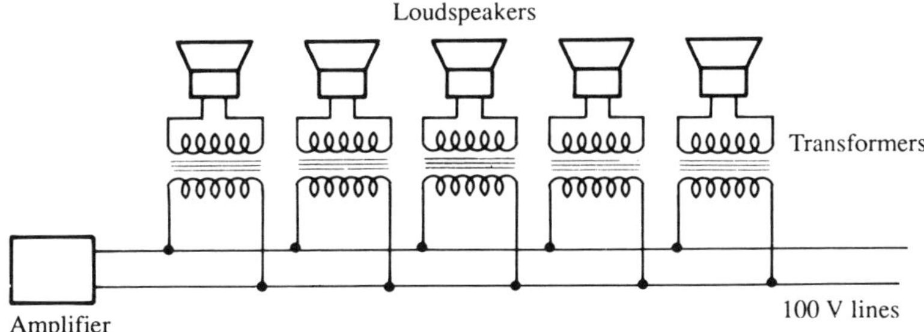

Figure 6.10 *Using the 100 V line system, which permits speakers to be run in parallel, with a separate transformer for each speaker to ensure correct transfer of power*

User guide

Users of PA and sound reinforcement systems need to be aware of the problems outlined above, because these strongly affect what can be done. In particular, microphone positions worked out by the AV technicians should not be arbitrarily altered, because it is likely these positions are optimum for both sound pickup and avoidance of feedback. Similarly, amplifier volume should not be increased to allow a greater distance between microphone and user. Alterations made should be made in conjunction with the AV staff.

The simplest use of a PA system is the speech-day type of presentations, where a single microphone is used. Such a system can be trimmed to offer good intelligibility even in an acoustically difficult hall, though a considerable amount of work may be needed. One point that is often overlooked is the absorbing effect of a large audience, so a hall which is unsatisfactorily resonant when empty is quite usable when an audience is present. This logically calls for some final trimming when the hall is occupied, something not always easy to arrange beforehand.

Musical concerts present more difficult decisions. A full orchestra or a brass band needs no artificial sound reinforcement, nor does a good piano, but some musical items, particularly faint voices or woodwind trios may need reinforcement. Only the minimum of reinforcement should be used for music, because acoustic musical instruments are, after all, designed to be audible in large spaces without amplification.

Staged plays are the most difficult challenge for sound reinforcement work. Amateur actors cannot be expected to project their voices like professionals. A single insensitive microphone cannot be used, because only a speaker close to it is picked up, so any solution must involve the use of several microphones. The usual method is to hang microphones from above the stage, positioning them so they are concealed by stage scenery. Feedback is the main problem, because these microphones need to be used at a fairly sensitive level to pick up all on-stage voices. Directional microphones need to be positioned so, between them, they cover the whole area of the stage otherwise dead spots will be noticeable.

A more expensive professional solution is the use of radio microphones carried by each player, or by a few key players. Radio microphones are, these days, accessible in terms of price, availability and licensing and form an excellent solution to all forms of sound reinforcement and sound recording problems. Construction of a radio microphone system is not beyond the skills of electronics enthusiasts, though approval may be needed to operate a system which is not a commercially obtained one, on the grounds of possible interference with other services.

Technical maintenance

Maintenance of public address, sound reinforcement and other sound communications equipment is fairly straightforward, and the main effort for the technician is setting up systems when required. Many halls have loudspeakers permanently mounted, with a PA amplifier located in a locked cupboard, so only microphones have to be provided and placed for the system to be used. Where less permanent arrangements have to be made, a suitable desk or other flat surface is needed for the amplifier and its control console. The control console should feature lockable controls to avoid alterations being made, which could cause feedback. One useful design feature for a main volume control is to split the action into two parts, one part of which is available only from within and is used merely to set range of the external control. This allows the external control to be fully used without risk of feedback, as its range has been limited.

Maintenance consists of cleaning, with particular emphasis on switch contacts and potentiometers. Noisy switch and potentiometer action is irritating and unacceptable, and reflects a low standard of maintenance. Switches on the system should be designed so operation does not create audible noise through loudspeakers. This implies that DC voltage levels on each side of the switched contacts must be equal at the time of switching.

Large amplifiers require dissipation of heat from output transistors (very occasionally valves), and heat-sinks must be kept clean and in a cooling airstream. If the amplifier, as often happens, is located in a hot little cupboard, an electric fan should be directed on to the heatsinks to ensure adequate cooling, particularly in summer months. However, most public buildings are grossly overheated (fuel conservation starts at home, and ends there) and such cooling may be needed just as urgently in the winter.

Microphones should be inspected at regular intervals, ensuring that connecting plugs and sockets are clean and making good contact. Foam surrounds and mesh grilles can become surprisingly dirty, and need to be detached and cleaned. Microphones must be handled carefully if protective grilles are removed, because the exposed diaphragm of a microphone is very easily damaged. Microphone stands should be checked to ensure they are extended easily when required, with no tendency to stick.

Loudspeaker grilles need to be removed and vacuum-cleaned at intervals. These grilles are made from acoustically transparent materials, most of which are very expensive, and great care should be taken to avoid damaging the material. If a grille is damaged, it cannot be substituted with other materials like cloth veiling without considerably altering loudspeaker treble characteristics. While grilles are removed, loudspeaker cones can be inspected for any sign of tearing or cracking, which would require replacement of the unit.

A loudspeaker which produces a prominent buzzing when used is usually suffering from either loose internal diaphragms or a detached voice coil. The enclosure will have to be opened to check in either case, and repairs made if internal diaphragms are loose. A detached voice coil cannot be repaired satisfactorily except by the manufacturers. Voice-coil damage is usually a sign the loud-speaker has been overloaded.

Hints and tips

- Keep microphones in their original cases when not in use.
- Loudspeakers need to be connected with heavy-gauge wire, as substantial currents are used.
- Microphone leads should always use screened wire. If low-impedance microphones (moving-coil or ribbon) are used, try to place the step-up transformer as close to the amplifier as possible as this reduces the chance of picking up interference.
- If PA equipment is old or has been home-made, it should be checked for electrical safety at intervals of not more than five years.
- If there is any structural alteration to a hall or to its furnishings, especially curtains, acoustics should be checked as soon as possible afterwards.

7

Photographic, copying and computer work

Photography has been a traditional skill of the AV technician for a considerable time, though not all AV centres are well-equipped for photographic work nowadays, particularly for enlarging except where this is taught as part of a course. Virtually no darkroom facilities are, however, needed for film development, and this includes the reversal processing of 35 mm (2″ by 2″) for slide projectors. The facility to prepare slides, along with a camera of the SLR class (now available at prices as low as £50), can make field trips considerably more interesting and allow the experiences of a field trip to be communicated to others later. The ability to prepare slides is also important for business presentations, as it requires less cumbersome equipment – every conference centre has a *Carousel* projector, few possess a PC computer with LCD projection screen.

The camera should be a 35 mm single lens reflex (SLR) type, such as Pentax, Nikon, Canon, Yashica, or a host of other manufacturers, not forgetting the very low cost Zenit from the USSR which provides the best ratio of performance to price obtainable anywhere. The dominance of the SLR has come about because of the advantage that what is seen in the viewfinder corresponds exactly to what is photographed, as the light path is through the lens (Figure 7.1). To this is added 'through the lens metering' (TTL) so exposure is metered or set automatically from the light that passes through the lens, as distinct from light simply reaching the camera or reaching a light-meter. The most recent development in SLR cameras has been automatic focus, using an infra-red beam system.

The main attraction of the SLR is the availability of interchangeable lenses to allow for any type of use from ultra wide-angle

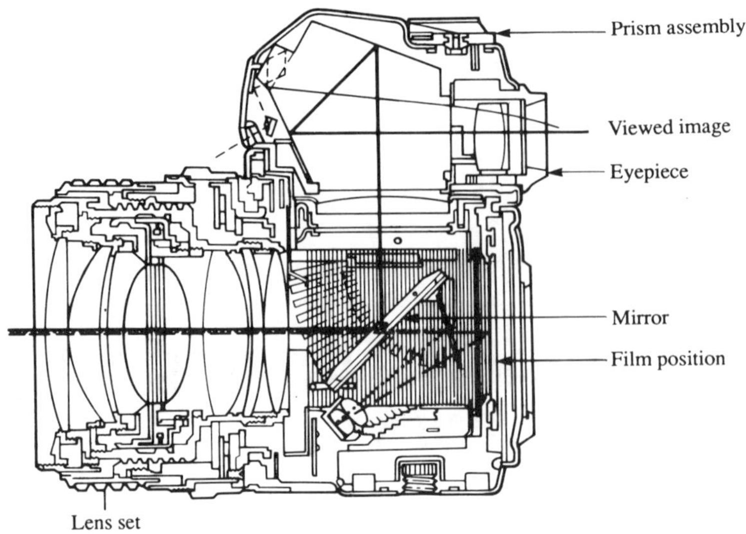

Figure 7.1   *Principles of the through-the-lens (TTL) reflex camera, showing only the main light path through which the image is seen for focusing. Part of this light is also used for metering. The mirror swings up just before the shutter opens to take a frame*

(fish-eye) to telescopic, and the adaptability to photomicrography which is particularly useful for biological work. Modern cameras have not consolidated on a single standard, though most use some version of a bayonet fitting for lenses. This is quicker to use than threaded lenses but there are still many cameras, particularly in lower price ranges, which use Pentax-thread lenses, and the range of low-cost lenses using the thread system is extensive. In particular, there are many excellent low-priced zoom lenses which can make it unnecessary to carry three separate lenses with the camera. Avoid, at all costs, being locked into a system which allows only the use of very costly lenses – though it is possible to use adapters so that other lenses can be fitted.

Microfiche readers

Where computers are not used for information storage it is common to find microfiche storage, with printed information reduced in size on to film transparency which can be read by a back-projection system. A typical microfiche reader, such as the

Bell & Howell *ABR–1055*, accepts all four standards of film data storage – jackets, microfiche itself, COM film and aperture cards. The principles are very similar to those of a back-projection slide system, but the film is placed on a 4″ by 6″ carrier which moves in both horizontal directions showing magnified views on the screen of different parts of the data contained on the film.

Modern readers use the same double-bulb technology as OHPs and slide projectors, so bulb failure (which is quite unusual) requires only the spare bulb to be switched into place, with the defective bulb replaced later when the machine is not being used. Screens are non-reflecting blue, grey or green tinted, and a 14.5 V 90 W bulb provides all the light intensity required because of the efficiency of back-projection. Controls consist only of focus and carrier movement, while the lens system is arranged to keep the film in focus over the whole area of the carrier travel, and to select high or low magnification without changing focus settings. Figure 7.2 shows the light path within the reader unit – note the number of mirrors used.

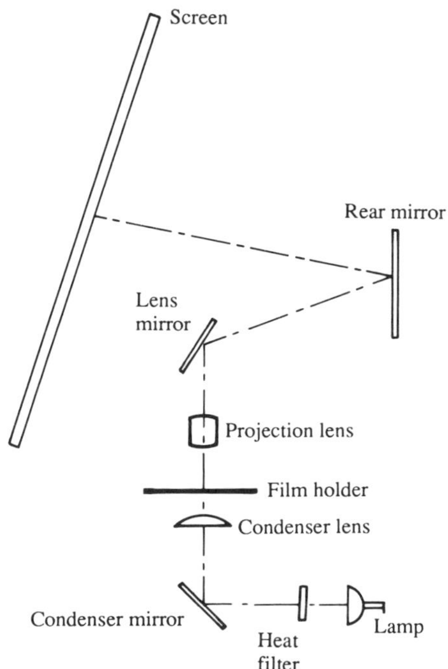

Figure 7.2   *Light path in a typical Bell & Howell microfiche reader*

Reprographics

The photocopier has now become another item in the care of the AV technicians, though a distinction is usually made between an office photocopier and the machine acquired for AV uses. The AV copier is used for preparing hand-outs for small scale use, masters for offset-lithographic copiers, transparencies for overhead projectors and so on, and should preferably be a type permitting changes of scale, both reduction and enlargement. Ideally, the AV photocopier should be a full-colour type which can be used to make good reproductions of photographs, something many office copiers do not do well.

Traditionally, the mainstay of reprographics work in the care of the AV department was the spirit duplicator. This cheap and sometimes effective method of copying suitable masters is still in use, but for higher quality work, with longer print runs, the offset litho machine has been added to the AV inventory. This creates more work for the AV department in the form of preparing masters for the offset machines, usually the cheaper foil masters.

Many AV departments now find themselves responsible for care and maintenance of computers and peripherals. Printers in particular are often held to be the responsibility of the AV section even if the remainder of the machinery is in the care of another department.

Principles – photographic work

Photography is one of the oldest of the AV technologies, dating to around 1835, though the principles had been established much earlier. Later developments have greatly refined camera design and techniques but have not greatly altered the basis of photography, which is to allow a light image to fall on to a surface of silver halide (bromide, chloride or iodide of silver, or a mixture) held in gelatine. The effect of light on this material is to cause separation of metallic silver to occur as a fine powder which looks black, so the image that is produced consists of black areas where the light image consisted of bright areas, making the image into a negative. Released halogen is absorbed into the gelatine so preventing the action from reversing. Unaffected or partly affected silver halide is then removed, using a material known as the *fixer* because it ensures the image would remain fixed on the backing material when it is exposed to normal illumination.

The optical aspect of photography consists, at its simplest, of a lens and a light-tight box, and many of the lowest-priced cameras of today are little more than this. At the other end of the range, computer-controlled production of lenses now makes it possible for low-cost lenses to approach standards formerly attainable only at very great effort and cost, and cameras themselves have achieved a high degree of automation. Technology, at the time of writing, is still based on the chemistry of silver halides, and it is likely to be changes in this aspect of photography which will bring the next set of changes in cameras and images. It is already possibly to buy a camera whose images are held on a magnetic disk or in a memory chip.

It is not necessary to expose the photographic material until an image appears, because even a comparatively brief exposure creates a latent invisible image, which is made visible with a chemical developer solution to complete the separation process. The developer is, chemically, a weak reducing agent which cannot separate silver from halogen by itself, but can finish the process that is started by the action of light. The developed image is then fixed in the usual way, with the material traditionally used – sodium thiosulphate – known by generations of photographers as *hypo* from its older chemical name.

Early processes used a reversal system, in which the final image is obtained the correct way round (not a negative) by chemical processing, but this is superseded by the negative-positive method used for many films today. However, reversal processes persist for slides and instant photographic processes. Its disadvantage of not permitting extra copies to be made is not quite so important in these days of easy copying techniques.

Colour photography has a surprisingly long history, and the principles were known even before black and white photography became commonplace. A colour film consists of three layers, one for each of the primary colours of light, with a layer of coloured filtering material for each sensitive layer. Sensitive layers are of ordinary silver halide material, but due to coloured filtering layers, separate layers of silver halide respond selectively to the colours in the image. Process theory dates back to the start of the nineteenth century, and all that was ever needed was the technology to make it possible. This was achieved by the start of the twentieth century, with considerable cost and difficulty, and colour photography for the masses, first using slides, later using print film, arrived in

Europe following the second world war, though it had been in use in the USA before this.

Processing of colour slides is remarkably simple, requiring no darkroom, and the method is outlined later in this chapter. Colour prints require considerably more effort and a high-quality enlarger, and it is possible now to buy miniature processing machines which carry out a complete cycle of development and printing. Black and white work should not be neglected, because good monochrome film produces high resolution with remarkably fine rendering of grey shades, and a day spent at an exhibition of first-class amateur (in the sense of unpaid) photographs will convince even the sceptic that colour is by no means the ultimate in photography. Good black and white work demands that the photographer should process his/her own work, and demands skilled use of the enlarger and the processing actions.

Following processing, slides are prepared by inserting film into card or plastic mounts; plastic type is preferred for use with *Carousel* projectors. Paper prints can be left as they are if matte grades of paper are used, but glossy paper is glazed by heating it while clamped against a polished metal surface (with a separating solution applied to prevent the print from sticking to the hot surface). Really good quality glazing is difficult to achieve, and many photographers prefer matte surfaces which do not cause distracting reflections in bright light.

Principles – copying machines

The spirit duplicator (Figure 7.3), uses a master prepared with a solid wax-ink coated sheet placed behind the paper, ink to the back of the paper. Typing, writing or drawing on the paper master then leaves a wax impression on the rear, used to print copies by moistening the master with spirit so the wax transfers. Dot matrix or daisywheel computer printers make acceptable masters, provided the impact is sufficient – dot matrix printers usually have adjustable impact force.

The main problem with spirit duplicators is to achieve correct moistening with spirit. The usual method consists of a felt pad fed from a reservoir that, in turn, is supplied from a tank with a pump. The pad moistens a roller pressed against the master copy, leaving the wax-ink moist. Over-enthusiastic pumping will flood the machine, wetting the master and preventing operation. This

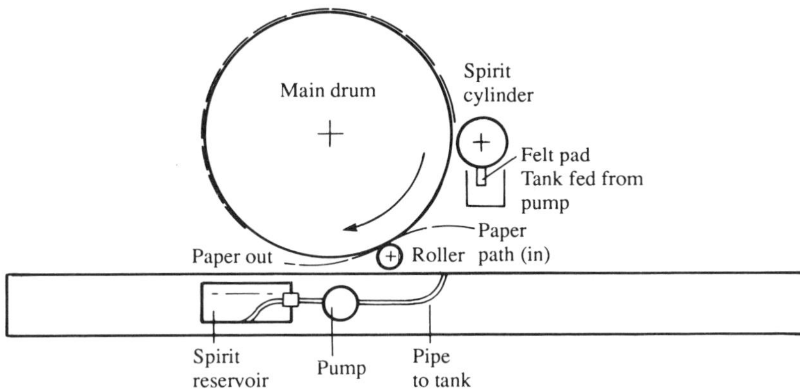

Figure 7.3   *Principles of the spirit duplicator*

accounts for the reluctance of anyone to use the machine on Monday morning or following a holiday.

The rationale of the spirit duplicator is that, in theory at least, it is cheap to run. Its main reason for survival is the last-minute hand-out, hastily written and duplicated and its use is steadily declining, particularly in places where computer systems with word processing programs are established.

Litho offset duplicating

Principles of offset lithography are ancient, as the name suggests. A surface can be wetted with water or with oil, but where water soaks in oil does not, and where oil is present water cannot reach, making these areas mutually exclusive. By making a pattern of wet zones on a surface, ink rolled over it adheres only to the parts which are not wet, and paper can be pressed against the surface to transfer the ink image.

Modern methods of lithography use a rotary offset principle, meaning a master plate is placed on a cylinder and never touches the paper to which its image is transferred. The master plate is made of metal or paper, and its ink image is rolled on to a resilient surface, usually a plate of synthetic rubber (known as a blanket) laid on another cylinder. This in turn transfers ink to the paper by rolling and pressing against each sheet. Enough ink must be delivered to the master to keep copies well inked, and enough water (usually modified by a solution supplied by the manufacturers) to dampen parts of the plate not inked.

Masters can be prepared on a typewriter, or a computer printer with a suitable ribbon, with a photocopier (the most popular method) or by direct drawing, and can be on metal plates for long master life or paper for shorter runs. Masters, whatever type, carry an image that is inked in a greasy material and when the machine is running both ink and the water solution are delivered in metered quantities. Relative amounts are critical, because a shortage of solution dries out the master and so more of it is ink-coated. Too much solution, on the other hand, makes ink adhesion poor, yielding thin and poorly printed copies.

Figure 7.4 shows a side view of the Rotaprint *R30/95*, a large

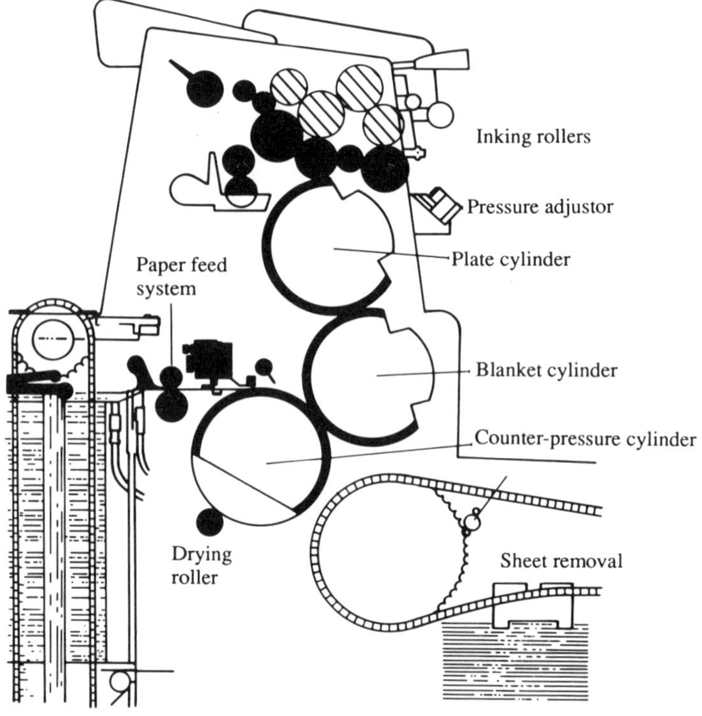

Figure 7.4   *A side view of a large offset-litho printer, showing the three main cylinders, plate, blanket and counter-pressure cylinders*

machine capable of printing up to 6000 copies per hour. A number of small rollers are used as part of the inking process, and are necessarily elaborate to ensure an even ink coating on the master plate on the plate cylinder. Two of these rollers oscillate from side to side to improve the uniformity of inking. Water solution is also transferred by these rollers of which one, the plate roller, is placed into contact with the plate cylinder when printing is taking place.

As the plate cylinder rotates, it picks up ink and solution, and an ink image is transferred from the master to the blanket carried on the blanket cylinder. Note that because the image is transferred to this intermediate roller, the original image is the correct (readable) way round. As the blanket cylinder rotates it also presses against the paper which is fed between blanket cylinder and counter-pressure cylinder. A drying roller under the counter-pressure cylinder ensures it does not become wet, as wet paper makes inking uncertain.

The main problem of all offset machines is in obtaining correct consistency of ink for the work to be done, and ensuring uniformity of inking. For many purposes, work is almost unchanging – reproduction of text and images on A4 paper – so that the ink consistency, once found, is easily maintained. Evenness of inking is dealt with by applying ink through a number of ducts (typically 25) each individually adjustable, across rollers.

The printing quality of an offset machine is so much better than a spirit duplicator's that these machines are almost universally preferred. Considerable care is needed, however, in keeping rollers clean and inking ducts correctly adjusted. Most offset machines can print in colour, by passing paper through again for each different ink colour to be used. This demands high accuracy of paper feed and location, and offset machines often use mechanisms such as suction feed, in which each sheet of paper is picked up on a suction pipe and placed precisely before being caught by the rollers.

Xerographic copying

The principles of the xerographic copier are different from those of any other printer, and AV technicians are often reluctant to carry out any kind of work on them. The printer uses an electrically charged drum (charged by an electrical discharge or *corona* which produces ozone as a by-product). This charge is removed as the material is rendered conductive when struck by a light beam. As the drum rotates, a light beam is scanned across the page to be copied and reflected light from the page is scanned by moving the light assembly along with mirrors and lenses across the page as shown in Figure 7.5.

Once this scanning process is complete, the drum contains an electrical voltage 'image' corresponding to the pattern of ink on the

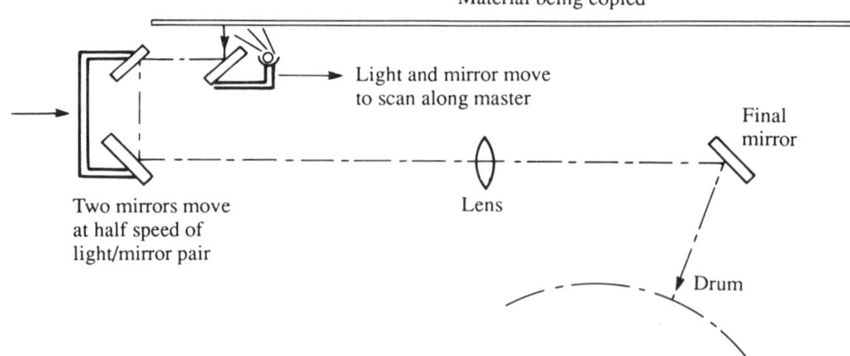

Figure 7.5   *How a master is scanned in a photocopier – the light and mirror assembly moves under the master, and two mirrors reflecting the beam to the lens move also, but at half the speed*

material copied – typically 100 V for white areas, 900 V for dark areas, and intermediate voltages for half-tone regions. Finely powdered resin, 'toner' is coated over the drum, sticking only where there is still an electric charge. The toner coating process is done using another roller, the developing cylinder, which is in contact with the toner powder – a form of dry ink. The toner is a magnetic powder, and the developing cylinder is magnetized to ensure it will be coated with toner as it revolves in contact with the toner cartridge. A scraper blade scrapes off excess toner to ensure the coating is even. As the developing cylinder rolls close to the main drum, toner is attracted across where the drum is electrically charged. Note that two forms of attraction, electrostatic and magnetic, are being used here.

Rolling paper over the drum will now pass the toner to the paper, using another corona discharge to attract the toner particles to the paper by placing a positive charge on it. After the toner transfer, the charge on the paper has to be neutralized to prevent the paper from remaining wrapped round the drum and this is done by a static-eliminator blade.

This leaves the toner only faintly adhering to the paper, and it needs to be fixed permanently into place. This is done by passing the paper between hot rollers which melt the toner onto the paper, giving the glossy appearance that is the mark of a good photocopier. Finally, the drum is then cleared of any residual toner by a sweeping blade, re-charged and made ready for the next page. Figure 7.6 illustrates this processes.

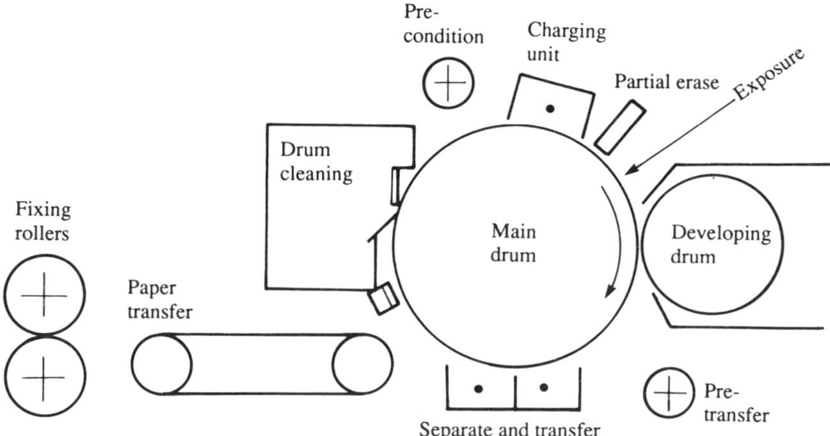

Figure 7.6   *Principles of the electrostatic copier*

The main consumables of this process are the toner and the drum. The toner for most modern copiers is contained in a replaceable cartridge, avoiding the need to decant this very fine powder from one container to another. The resin is comparatively harmless, but all fine powders are a risk both to lungs and in terms of explosion danger. The drum is usually coated with selenium, which should not be handled and which gives off toxic gases if it is ignited (selenium is a close relative of sulphur, and is flammable). *Never open a photocopier which is operating.* Some copiers use a less objectionable zinc oxide coating, and a few now use organic photoconductors (OPC) which are of low toxicity and which do not have to be returned to the manufacturer. Selenium and zinc-oxide coated drums must be returned to the manufacturer after replacement. Drum replacement is, on average, needed after each 80,000 copies, and less major maintenance after every 20,000 copies.

Principles – computer printers

Printers used with small computers use one of the mechanisms listed in Table 7.1. Of these, impact dot matrix types are the most common and versatile, though their predominance is challenged by lower-priced Xerographic types (LED and LCD), ink-jet dot matrix types, and thermal wax-ribbon dot matrix types.

A dot matrix printer creates each character out of a set of dots – when you look closely at the print you can see the dot structure.

Table 7.1   Types of printers available

Types	*Classification*
Dot matrix	Inked ribbon, impact Waxed ribbon, thermal Sensitive paper, thermal Sensitive paper, spark Inkjet, spray
Xerographic	Laser beam scan page LED array line LCD mask line
Type impact	Lineprinters Daisywheel

Most dot matrix printers are impact types: paper is marked by the impact of a needle on an inked ribbon which hits the paper.

Impact dot matrix printers use a printhead containing 9, 18 or 24 pins (sometimes called wires or needles). The mechanism for each pin is quite elaborate, and the principle is illustrated in Figure 7.7. Each

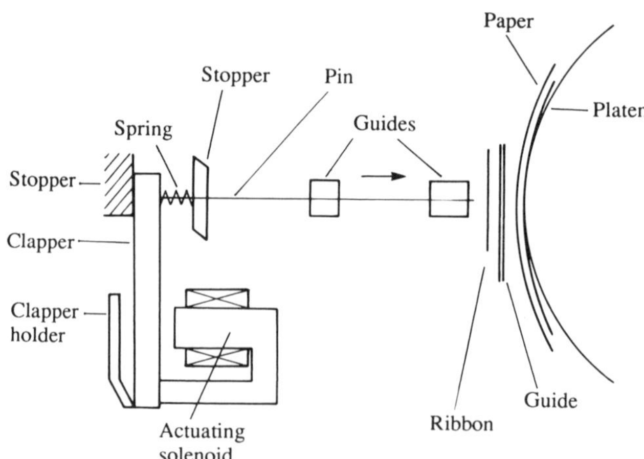

Figure 7.7   *A typical dot-matrix printer mechanism. This is miniaturized so that 9, 24 or even 48 pins can be located on a rectangle the size of a printed character.*

pin is held in guides, and when the actuating solenoid (an electromagnet) is activated, a clapper strikes the pin, driving it against the ribbon and paper to produce a dot. The remarkable aspect of a dot-matrix printer is that this mechanism can be miniaturized to such an extent as to pack so many pins into an area the size of a printed character, about 3 mm × 2 mm. In fact, only the ends of the pins are packed so closely, so pins are flexible enough to allow the rest of the mechanism to be more widely spaced.

Generally 9-pin printers are most common and so are cheapest (although some are sold at very high prices because of their particularly robust construction or high-speed printing or both) but the trend is to 24-pin printers. More pins means a higher quality of print. By using two slightly staggered vertical rows of 12 pins each (Figure 7.8) these printers can print at a high speed and with excellent quality with little of the 'dotty' appearance associated with 9-pin dot matrix printers. Most 24-pin printers have built-in fonts so a range of type-styles are selectable by switches on the printer itself, or under software control from your word-processor. For general-purpose use a 24-pin impact dot matrix printer using its own fonts is very cost-effective.

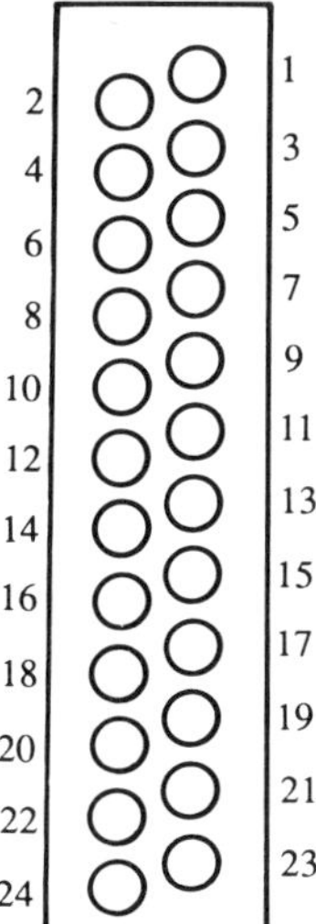

Figure 7.8   *Arrangement of pins in a 24-pin printer*

Impact dot matrix printers can be used with carbon or NCR (no carbon required) paper to produce duplicates, not possible with non-impact printing methods. If you want two copies of a document from an ink-jet or laser printer, you have to separately print two copies.

The other main variant on the dot matrix theme is the ink-jet type, which uses a matrix of tiny needles to squirt ink at the paper.

In the past a single jet, electrostatically moved was common, but the more usual mechanism for modern inkjet types is the bubble-jet principle, developed by Canon. In this, jets are filled with ink at all times, and ink is expelled out by rapidly heating a needle so a bubble forms, blowing a drop of ink out. Application of this principle has resulted in a new generation of silent printers producing excellent character shapes with virtually no trace of dot pattern, comparable to the output of a good laser printer. Though prices of high-quality ink-jet machines have dropped, they remain costly machines to use because the ink reservoir and jet system is in one piece, renewed each time the ink runs out. This ensures there are few problems with clogged jets, but running costs are quite high; comparable with those of a laser printer.

The ultimate in print quality – for general-purpose use at least – is provided by laser printers, which include variants such as LED-bar and LCD-mask printers. These base their technology on xerographic copiers, and are quite fast and silent in action. Laser types are often page-printers, meaning it is necessary to store a complete page of information in printer memory to print the page. Quoted speeds of most laser printers refer to repeated copies of a single page, not to normal printing of single copies, which is considerably slower.

The laser printer is basically a copier in which the drum is scanned by a laser beam, which has the effect of making the drum material conductive. The beam is controlled by a pattern of signals held in printer memory and enough memory must be present to store information for a complete page. This requires about 0.5Mb as a minimum for text work, and 2Mb or more if elaborate high-resolution graphics patterns are to be printed. The remainder of the action is identical to an electrostatic photocopier.

LED and LCD printers work a line at a time, so they do not need the huge memory required by laser printers to store a complete page image. In place of the mirror or prism and laser arrangement, these printers use a line of miniature light-emitting diodes (LEDs), or LCD cells with a backlight, controlled by signals making up a line of dots in a line of print – if there are 80 characters per line on a 9×14 matrix, there are $9 \times 80 = 720$ diodes or cells, and the paper is moved horizontally upwards fourteen times to produce a complete line of print. In all other respects, the mechanism is identical to the laser printer's.

An older technology is represented by the daisy-wheel printer. This uses a typewriter approach, with letters and other characters placed on stalks round a wheel. The principle is that the wheel spins so the letter you want is at the top, then a small hammer hits the back of the letter, pressing it against ribbon and paper. The quality of print is high, but machines are noisy and slow. They have been almost entirely replaced by 24-pin dot matrix and laser types for most business and educational applications, though they can still be found (and heard) in some smaller offices. Their lack of versatility has been the deciding factor in their decline – changing a font is done by changing the print-wheel, and no graphics can be printed. However, daisywheel printers can now be bought more cheaply than portable typewriters.

An entirely different approach to producing paper output, mainly for graphics, is the pen plotter, which uses a set of up to eight miniature pens moving over a sheet of paper. Top-quality plotters are expensive, but ideally suited to CAD work. Lower-cost plotters are available, though mostly for work on small paper sizes. Plotters are ideally suited to producing output in colour because each pen can be a different colour. Though they can produce text characters (writing each character rather than printing it) they are very slow compared with printers for this type of application. Pen plotters may be required if computer-aided design and drafting is taught, particularly in architectural studies.

Connectors

Printers for all but a few computers almost universally use a parallel interface. Each character you see on screen is represented by a set of signals, usually seven but sometimes eight digital bits. Parallel transmission means that all eight signals from the compu-ter are sent together from computer to printer, along with synchronizing signals. The cable which connects computer and parallel printer therefore has eight separate strands for character signals plus several more strands for synchronizing signals. When a character is transmitted using this system, each data line carries one particular bit, so all bits of a character are sent simultaneously. Sychronizing, sometimes called handshaking, is required only to control the speed at which the computer sends the character signals so the printer is not overloaded. The system was originally developed by the printer manufacturer Centronics, and is often referred to as the Centronics parallel interface.

The main problems of parallel data transfer as far as printers are concerned are line length and rate at which signals are sent. Each signal is a very short, sharp change of electrical voltage, and the longer the length of cable, the more likely it is signals are corrupted. This can happen in two ways. First, short, sharp changes of voltage are rounded out into long smooth changes, ruining signal timing. Second, a signal on one line may be picked up on another line, causing an error at the printer end of the cable. The practical effect is to restrict parallel printer leads from computers to 1 to 2 metres. You can, if you need to, get around this restriction by using repeaters, amplifiers which restore correct signals at the end of a long line, but few PC owners take this way out of a cable length problem.

For many purposes, it is often satisfactory to use only one printer for a group of computers. One way of achieving this is to network all machines, but cost of networking is high. An attractive alternative may be a printer switch. Such switches allow several computers to be connected to one printer through manually-operated switch. Each computer user simply switches the printer to a desired computer prior to printing. After printing, the switch is altered to allow another computer to use the printer. Using, say, a four-way switch to allow four computers to share a laser printer is a very satisfactory method of sharing an expensive resource. Note: some types of printer switches are unsuitable for use with Hewlett Packard laserjet printers.

User guide – cameras

The traditional non-automatic SLR cameras demand you select a subject, a viewpoint and a suitable lens. The lens is focused on the subject, and the picture composed. Exposure is determined by setting shutter speed suitably (a short time of say, 1/250 second or less, if the subject is moving or if a heavy camera is being held in the hand rather than on a tripod), and the aperture of the lens is adjusted so that exposure is correct. The shot is then taken by slowly and gently pressing the shutter release until the shutter fires. The film is then wound on for another exposure.

Preparation of the camera may have been left to the technician, but it is wise to check the film type that is used. Film speed is indicated in ASA units, of which the most common for colour print use is 100 ASA, a comparatively slow film. Films with speeds up to 400 ASA are widely available (supermarkets, chemists and so on), but higher speed film will generally only be stocked by photographic specialists. Black and white film in very high speeds (of the order of 3200 ASA) can be obtained. Many black and white films may be used with faster shutter speeds than the film rating indicates if the development process can be adjusted to suit, but this is applicable only if you are developing your own films or if the film is developed professionally; it is not widely available commercially.

Recent cameras feature the *DX* automatic film speed coding system, in which a set of notches in the film canister convey the film speed information, and set the camera accordingly, dispensing with the setting dial. If an old film with no DX coding is used, it is exposed at 100 ASA.

Film in the camera has to be treated with some consideration, particularly with regard to temperature. Colour film, in particular, deteriorates quite rapidly at high temperatures and though only the keenest of users carry film around in refrigerated containers (as some do), it is foolish to leave a loaded camera on the parcel shelf of a car. Film shelf life is limited and so film should be exposed and processed within a reasonable time, say six months.

If an exposed film is accidentally dropped in water, it should be kept wet in clean water, and processed as soon as possible – this will probably have to be done in the AV darkroom as it is not acceptable to send wet film through the post.

User guide – microfiche readers

Microfiche should be kept covered until required, because any mark on film is greatly magnified when placed in the reader. Reading a microfiche involves inserting it in the carrier (pull the carrier towards you to open the top glass, place film into the carrier, and switch on the reader), focusing on to the film, then moving the carrier to find the part of film you want. Where microfiche readers have magnification settings, it is best to start at a low setting then, once you have located what you need, the higher magnification can be used for reading. Remove the film after use and replace it in its packing.

One important prohibition concerns writing on the screen. It may be a considerable temptation to locate data by marking its screen position with a ball-point or felt-tipped pen. Screen surface, however, is matte and such marks are very difficult, often impossible to remove. The only satisfactory treatment is replacement of the screen.

User guide – reprographics

Spirit duplicators should be avoided except in desperate circumstances (such as having no other duplicator), but if they must be used a few hints can make experiences more bearable. Paper should be fairly glossy, but it is useful to have some absorbent paper for preparing the machine. This involves pumping up spirit until the machine is wet, and putting a sheet of plain absorbent paper in place of a master. Roll this around a few times, with no other paper going through, to absorb excess moisture, and do not put in the master until you are convinced there is spirit present (your nose is the most reliable indicator) and flooding is not apparent.

Masters should be fresh, if possible, and well cut. Put the master on to the drum, carbon-wax printed side out, and place glossy paper into the pick-up tray (look at this paper carefully, and if one side is more glossy than the other, use the more glossy side upwards). Crank the handle to roll in the first sheet. If this is evenly printed, be grateful, and print off the rest. Throw the master away unless you expect to print another batch in the next few days.

A variation on this theme is to make all masters using a computer and daisywheel printer. Daisywheel printers make superb spirit masters, because they hit the paper very hard, yet print a page in a reasonable time. Further, using a computer allows you to keep a large supply of handouts on a single 3.5" disk, making copies as and when needed.

User guide – offset litho

It is unusual for the offset litho machine to be made available directly to casual users, because experience is needed to obtain good results. Some machines, however, have reached a fairly automated state and can be set up by the technician so the user need only know how to attach a plate, ink up, and switch on. Before starting a session, the machine should be loaded with enough paper for the job in hand.

Typically, the master is held upside down and hooked on to the head-plate clip. The machine is then slowly turned round by hand (with the ink rollers pulled back and no pressure applied to the other rollers) until the master has been fed in without wrinkling, and can be attached to the tail-clip. A metal plate may have to be wiped with a solution to remove preserving solution; a paper plate needs to be moistened with a different solution. As plates soon dry out, the machine should be started as soon as possible and ink level moved to its normal running position. Set the number of copies required so the machine runs until the required number are printed.

Cleaning is carried out by the technician and is not left to the casual user. All the rollers need to be cleaned and the ink duct and water fount emptied, cleaned and re-filled. This is normally done at the end of a day's printing, though for machines where a comparatively small number of copies are printed each day this action is often carried out once a week only.

User guide – computer printers

If computers are used in single stations, where each computer has its own printer and disk drive, it is a simple task of first selecting a computer which has the type of printer (daisywheel, impact dot matrix or laser) you need for your application. One problem is that many schools and colleges have a range of non-compatible types, so that the disk you use for your text may not be readable by all computers. This is less of a problem if the computers are IBM-compatible.

A common solution to the resources problem is the networking of computers, so any one of a set of connected computers can be used to operate a printer which all of them share. When the printer is available, check the ribbon is in a suitable state for your use. If the ribbon is worn it should be replaced or, more likely, recycled – hand the ribbon to the AV technician for re-inking and receive in exchange one that has been freshly inked. Daisywheel machines are likely to use carbon ribbons which have a reasonably long life, and as the ribbon is visible it is not difficult to estimate whether it will last long enough for your work. Laser printers require only an adequate supply of toner, and there should be a warning system if the level drops too low.

A more difficult check is to ensure the printer will do what is required. Dot matrix and laser printers in particular can be set up to various emulations and have large numbers of optional settings. You might, for example, want to see zeros slashed or unslashed, or reproduce some graphics characters (such as boxes around text, or the English pound sign). Do not take it for granted a printer will print what is visible on the screen. Further, a previous user may leave the printer in a state unsuitable for your purposes. If the technician has a program that sets up the printer as you require it, put it on your disks and run it prior to all your printing sessions.

Printing text or graphics usually requires suitable software to be present and running, so that you might need to carry a program disk or to load a program from the network. For text alone, a word-processor program is generally used, but a mixture of text and graphics often calls for a desk-top publishing (DTP) program. Such programs often require the software to be set up for a particular printer, and a printer-setup program may have to be run before using the main software. After printing, changes made to printer settings should either be reversed, or the technician advised. Where printer consumables are expensive, as for a laser printer, you may be required to log the number of sheets printed. Once set up to print a page, laser printers can make repeat copies very quickly, so the use of a laser printer avoids the need to use the photocopier or the offset, and this method of preparing text and graphics is gaining favour as a result.

Technical maintenance – cameras and darkroom

Cameras should be cleaned each time a film is removed, and the most useful instrument for cleaning is a brush fitted with an air-bulb. The combined action of brushing and puffing removes most loose particles which otherwise could settle on film and cause marks. Lenses should be lightly polished – the key word is *lightly* – many old cameras bear circular marks on lenses where enthusiastic but misguided rubbing has scratched surfaces. Film-wind and shutter actions should be checked, and if a camera is faulty it should be returned to the manufacturers for servicing – few AV departments possess equipment or suitable dust-free room for a full service on a modern camera. Unless DX coding is in use, speed of the film loaded should be noted, usually by clipping part of the film box onto the holder provided on many SLR cameras.

Slide and possibly colour negative film can normally be processed by the technician. Details of processing will be contained with the processing chemical pack, but the following brief outline of reversal processing of colour slide film may be useful to the technician who is not well acquainted with the methods. Film which is suitable for non-commercial processing methods, such as *Ferraniacolor* or *Orwo* film or *Barfen* material, must be used. Such film can be bought in long (10 m to 30 m) rolls contained in tins, and loaded into cassettes by the technician.

Processing starts by preparing enough water to about 18°C to make up the chemicals. All colour processing depends on combinations of time and temperature and both must be reasonably precise; temperature to better than 0.5°C and time to the nearest second. A commonly-used method is to fill a large container (an old bath is ideal) with water at just above the correct temperature and to place all containers of solutions, suitably anchored, into this constant-temperature bath.

In the absence of a darkroom, a changing bag can be used for film loading and transfer of film into a developing tank. Changing bags comprise a double-layer bag with each layer zipped. Canned film, developing tank, and cassettes, as required, are placed in the bag through the main aperture, and both zips closed. By inserting arms through elasticated apertures users then work on film by touch. Film is transferred from can to cassette, or from cassette to developing tank, and cassette and can or developing tank sealed up again. Once this is done, arms are withdrawn from the bag and the bag is opened so all further processing is carried out in the light. It is always desirable to use subdued light in case light sealing of tanks or other devices is not perfect.

A developing (Figure 7.9) tank contains a spiral film holder into which film is loaded from a cassette. Doing this without seeing the

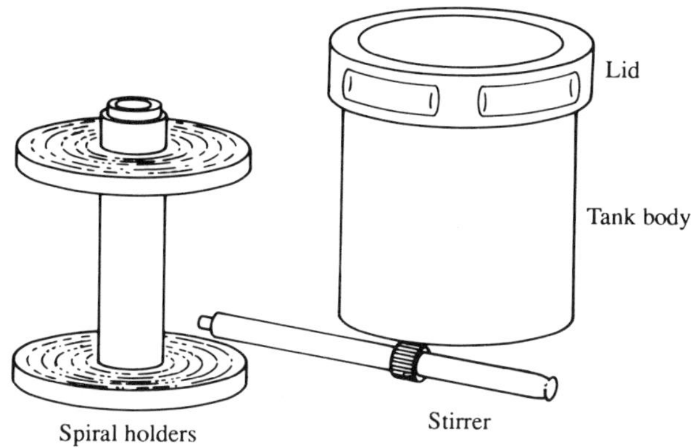

Figure 7.9 *A developing tank, showing the spiral into which the film is loaded in darkness. Once the film and spiral are enclosed in the tank, the rest of the processing can be carried out in the light*

film requires some practice, using an old reel of developed or scrap film. Practise first by looking at what you are doing; opening the developing tank, taking out the spiral, opening the cassette and feeding film from its reel into the tank. The difficult part is feeding the first piece of film into place, so the sprocket holes of the film engage, after which most spirals feed film automatically as you turn the two parts of the spiral alternately. Once you can load in a film as you are watching, practise with the scrap film using the changing bag. Only when you are convinced you can load a film without mishap should you try loading a tank spiral with a film for processing. One complication is that unprocessed film tends to stick in the spiral to a greater extent than scrap (because of the gelatine), and this is particularly noticeable in high-humidity conditions. This makes loading a slow process, because it is often possible to advance film by only a few millimetres at a time after about half the film is loaded. This is a good argument for using cassettes with short lengths of film, 10 to 12 exposures.

Once the spiral is loaded with film for processing, put it into the tank and secure the top – check that the top is correctly fastened. With the tank out of the bag, prepare the chemicals, preferably well away from the tank, or with the tank in a covered place.

For colour reversal processing you need a first developer, colour developer, bleach and fixer, along with clean water rinses. A sink and running water supply is best, though it is possible to work in a restricted way with just a large water-bottle. Do remember, though, that photographic processing requires a lot of water, and economizing on water results in slides of a short life. In addition to chemicals you need a light source to expose the film, usually a 60W bulb with no shade. You need also a receptacle for used chemical solutions – they should not be poured down the sink.

The process then comprises of checking the temperature of the first developer, pouring it into the tank to the correct level, and gently stirring the spiral to and fro with the stirrer, which engages into the inner core of the spiral. This is done for the specified time, and the tank is emptied then rinsed several times with clean water at about the same temperature as that of the other chemicals. This first development creates the set of black-and-white images.

The spiral is then taken out of the tank and exposed for a set time of a few minutes to the 60 W bulb (a few processing packages omit this step, because a chemical fogging process is used in its place). Following the second exposure, the film spiral is replaced in the tank and colour developer is poured in, once again noting temperature and agitating the spiral for the recommended time. If the temperature has changed since the solution was made up, time is changed accordingly – see the detailed instructions on the pack. At the end of colour development, the solution is poured out and the tank rinsed with several changes of clean water. In identical procedures bleach bath is used for the recommended time to remove the silver image, leaving the dyes created in the colour development, and after another rinse stage, fixer is poured followed by a final rinse. The tank is often rinsed simply by leaving it under a running tap, but this, though less labour-intensive, is not as effective as several complete changes of water, stirring well after each change.

The final rinse should, if possible, be in demineralized or distilled water to which a drop of wetting agent is added. If only hard tap water is available, follow the manufacturer's recommendations on use of wetting agent. Note that water from a water-softener of the ion-exchange type stains the film just as badly as hard water, though stains can be washed away in distilled water later. The film is hung up to dry in a dust-free space – never attempt to dry films rapidly in front of an electric fire or fan-heater.

Black and white film can also be reversal processed, using chemicals that can be made up fairly easily; and colour or monochrome developing and fixing for print films is as easy to carry out, and less critical of temperature and time. One important point is that most photographic chemicals are hazardous. Colour developers in particular are skin irritants and like all dyestuffs are related to carcinogenic materials – always wear gloves while handling these solutions, and dispose of them appropriately. Contact your local water board for advice about disposal of photographic chemicals; schools and colleges usually treat these chemicals to reduce toxicity before disposal.

Technical maintenance – enlarging

Where a good darkroom exists, enlarging is a normal part of the AV technician's activities. Elements of black and white enlarging are learned fairly quickly, but to turn out consistently good work requires a considerable amount of care, knowledge, dedication and experience. If colour enlarging is done a good sense of colour is required, something surprisingly rare in the male of the species.

An enlarger is basically a low-powered slide projector arranged on end, so it projects an image down on to a horizontal surface

(Figure 7.10). The image can be focused and composed by selecting image parts, so when a print is made (by placing a sheet of enlarging paper on the baseboard) the picture need not necessarily be that of the complete negative. In addition, features can be emphasized or reduced by selective work during exposure in the enlarger and it is even possible to indulge in faking by merging parts of two negatives together. The negative is held in a carrier, so different negative sizes, are possible with different carriers.

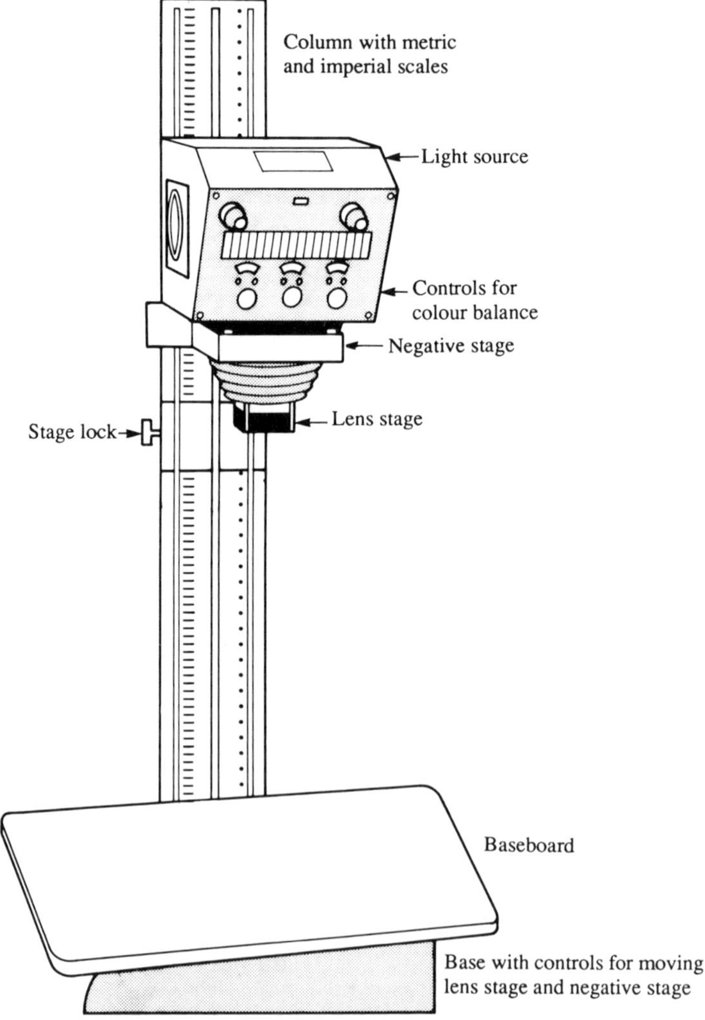

Figure 7.10　*Components of a modern enlarger. For professional work, the assembly is made from metal castings and is heavy and rigid, ensuring a steady image on the baseplate*

The light source can be the conventional quartz-halogen bulb and condenser, or a cold-cathode discharge source which gives better suppression of dust and scratchmarks.

Exposed paper is processed using the same developing and printing actions as required on the film negative. As usual, colour printing requires a considerably larger number of chemical baths than black and white, and control of temperature is more critical. Black and white print development can be carried out in subdued light, but this is not generally possible with colour printing, so an enclosed colour print processing unit is a better answer for this type of work. A process particularly well suited to the one-off printer is *Cibachrome*, obtained from any photographic suppliers specializing in amateur work. With care, *Cibachrome* gives results of exhibition standard and which even on a routine basis are better than most commercial colour prints. An enlarger for colour work requires a colour head containing filters so that the light of the enlarger lamp can be adapted to give a better balance to each negative. Colour analysers can be used to show what adjustments should be made to the colour head to obtain best colour rendering for each negative.

A good quality enlarger is essential for work of a semi-professional nature, and a darkroom used by students of photography should be equipped with enlargers, such as the De Vere range, which are sturdy and intended for professional use. Cheaper enlargers intended for budget purposes are generally too flimsy to stand up to intensive handling and will turn out to be disappointing in the long run.

Darkroom tidiness is important, and should be drilled into all users. In particular, chemicals should be properly disposed of, and no open packets left lying about. Dust from developer and fixer powders in particular have devastating effects on film and paper stocks, and the best way of avoiding this in a small darkroom is either to buy all chemicals as concentrated solutions, or to make up solutions outside the darkroom. All bottles used for transporting such chemicals should be prominently labelled, and arrangements must be made for correct and prompt disposal of used chemicals.

Technical maintenance – duplicators

Where spirit duplicators are in use, much of the AV technician's time is spent servicing these machines and mollifying users who are unable to make copies. Check level of spirit, renew the felt pads as soon as they show signs of wear, and keep an eye on the condition of the plastic feeder pipes – these often split so that spirit goes everywhere except into the felt pad. Ensure that gears are lubricated and turning freely and the paper grip on the drum is functioning. Most complaints, other than those concerning spirit apply to the paper-feed mechanism, either it refuses to pick up paper or it grabs too many sheets at once. Show staff how to shuffle a pack of paper to make it easier to feed correctly, and make certain the pickup roller has a good tread on it – these can wear smooth in a surprisingly short time. A wood-rasp is useful for re-treading pickup wheels, but as a temporary measure a rubber band wound around the wheel works wonders. Another useful treatment is a coating of rubber solution such as Evo-stik.

The offset litho printer requires attention as described in its operator's handbook – anything more than is covered in this handbook should be dealt with by the manufacturer. Most frequently required maintenance is cleaning of all the inking rollers, of which there may be a large number (Figure 7.11). Most schools and colleges are likely to use the smaller type of office

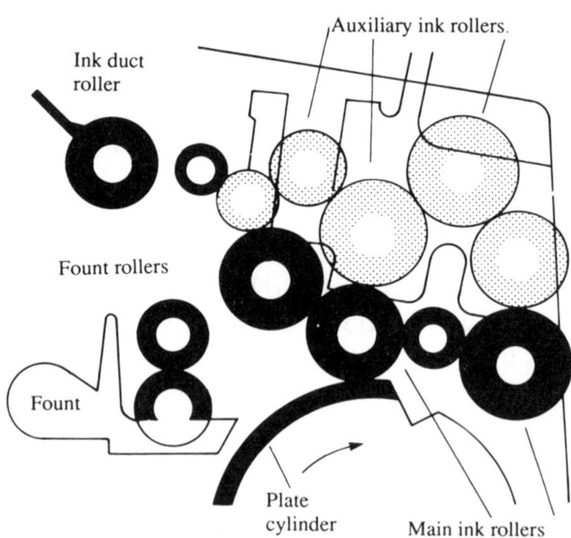

Figure 7.11   *The inking rollers of an offset-litho machine, which must be cleaned after use*

machines, handling A4 or smaller, but a few more specialized AV centres may use larger commercial sizes handling paper up to 13.5′ × 20′. Larger machines have extensive lubrication schedules with typically 50 or more lubrication points, some of which need attention at intervals as short as 8 hours, others at 40 hours, some only as required. Smaller machines, particularly those designed for office use, often feature plastic bearings requiring no lubrication. Larger machines use synthetic blankets attached to blanket cylinders and such a blanket should be thoroughly cleaned at the end of each print run – a spare blanket allows one run to be made after another. On smaller machines, blankets and blanket cylinders are not readily separated, and the cylinder must be cleaned. In some cases this can be done without removing the cylinder, in other machines the blanket cylinder has to be removed. Cylinder removal should not be attempted on large offset machines unless the full maintenance manual is available – apart from anything else, the weight of cylinders in a large offset machine is considerable.

Technical maintenance – computer printers

Care of computer printers covers both mechanisms and ribbons, of which ribbons require much more attention. Older daisywheel printers which use the IBM type of carbon ribbon often allow a long ribbon life, and as the ribbon is visible it is not difficult to decide when it needs to be changed. Ribbons for dot matrix printers vary considerably in type and life. Those housed in a long casing and which remain stationary as the print head moves generally have a long working life, and seldom need re-inking or replacement. The exception occurs if such ribbon casings contain only a short piece of ribbon, in which case, change your supplier. Smaller types which move with the printhead, and particularly those used on 24-pin printers, have a short life, often requiring changing in the middle of a large document. To re-ink such a ribbon:

- Open the casing carefully, pulling back any clips and using a screwdriver blade as a wedge to part dowel joints. Be careful not to lose or misplace any of the small gearwheels and springs inside the casing.
- Lay the lower part of the casing, which contains the packed ribbon, down on an open newspaper. Spray with *Caspell* ink spray or, if not available, with *WD40* silicone lubricant. Spray sparingly – do not make the ribbon visibly wet.
- Reassemble the top of the casing, making sure no gearwheels or springs are displaced and the ribbon path is not changed.

Most casings use a ribbon path which reverses the ribbon side in each run through the ribbon.

- Clamp the casing back in place – fastenings usually remain tight even after many re-inkings, but sticky tape can be used to hold the casing halves together if necessary.
- Wind on the ribbon to check that it runs smoothly through the casing.

Re-inking can, with advantage, be done even on new ribbons, because such ribbons often dry out in storage and produce poor results. Avoid over-inking at all costs, because it can take a long time to dry out an over-inked ribbon, and a lot of printing can be ruined in the meantime. It is better to store a re-inked ribbon for several days before using it, wrapping the ribbon in its original packing to prevent it from drying out too rapidly.

Make sure that you know the position and settings of miniature switches (dip-switches) within the printer, and keep a note of the effects of these settings stuck either inside the printer casing or under the printer. At monthly intervals, remove printer rollers and clean them thoroughly. If there is a lubrication schedule for the printer, observe the recommendations. Always be sparing with lubrication, and avoid using mineral oils on plastics.

Laser printers require the usual consumables of paper and toner, also requiring periodic replacement of various parts. The service manual details this schedule, emphasizing parts to be replaced at appropriate intervals (usually 100,000 prints) whether old parts show any signs of wear or not. Parts usually include fixing assembly, pick-up roller, separation pad, transfer corona assembly and ozone filter.

Most maintenance is needed when toner is replaced. Many laser printers use a cartridge (EPS cartridge) containing both toner and photosensitive drum, and the whole cartridge is replaced at recommended intervals. At this time it is normal to clean the fixing assembly and its cleaning felt, transfer corona assembly, primary corona assembly and transfer guide. Refer to the laser printer technical manual for a full description of maintenance schedules and for details of fault diagnosis.

Technical maintenace – microfiche readers

Microfiche readers are very reliable and need little maintenance other than periodic cleaning. One point is ball-pen and felt-tipped marks on the screen are very difficult to remove, and the screen may have to be renewed if marking is serious. The screen should be cleaned with glass-cleaner and a soft cloth, and lens and mirror surfaces cleaned with alcohol and soft tissues, preferably *Kleenex* facial tissue. Be careful not to spill alcohol or other solvents, because optical black paint used to prevent unwanted reflections within a reader dissolves in solvent. Large particles should be picked off rather than rubbed away, and any trace of adhesive from sticky tape should be removed using nail-varnish remover (amyl acetate) or toluene on cotton-wool buds. Any trace of fibres should be blown off after surfaces dry.

Check optical performance after cleaning. Focus should be sharp and clear and any faults here are likely to be due to dirt on a lens or an incorrectly located lens. Screen shadows are due to badly-adjusted mirrors or incorrect positioning of transformer-lamp assembly. Poor overall screen illumination points to a lamp near the end of its life, or dirty lenses and mirrors.

Hints and tips

- The *Manual of Photography*, published by Focal Press (formerly the Ilford Manual of Photography) is the most comprehensive sourcebook available, and contains a wealth of useful advice for photographers at any level.
- Materials and equipment for photography can be found advertised in the magazine Amateur Photographer. One supplier in particular, Eric Fishwick (see addresses at the end of the book) maintains an enormous stock and is very helpful for AV photographic equipment (including overhead projectors, slide projectors, enlargers and darkroom equipment).
- Manufacturers of cameras and other photographic equipment are usually helpful in providing educational aids such as cut-away cross sections, diagrams of optical paths, processing guides and so on. This can be useful for both staff training and student use.
- When a felt pad for a spirit duplicator is renewed, it is helpful to soak it in a 50% alcohol/water mixture, then dry it out partially before fitting it. This make it easier to wet the pad with spirit when it is put into service.
- At the start of a holiday period, remove all felt pads from spirit duplicators, soak them in spirit, and put them into a sealed jar.

- Put a cover over all photocopiers at the start of a holiday period to prevent dust marks on copied material later.
- Remove ribbons and cover computer printers at the start of any holiday.
- Periodically check ventilation in a room containing a photocopier or laser printer. These machines generate ozone in their electrostatic processes and this needs to be removed. A significant concentration of ozone has odd effects such as perishing rubber bands and is not (whatever the old seaside posters used to say) good for human health.
- Ink on offset machines dries up completely during an idle period, so the ink rollers and ink supply pipes should be cleaned before a holiday break and left wrapped to avoid gathering dust. If the blanket is removeable it should also be cleaned and packed away, and the whole machine should be covered.
- Ventilation in darkrooms also needs attention, because several types of photographic chemicals vaporize readily.

Glossary

ASA A system of measuring film speed (sensitivity) in linear units – doubling the ASA number corresponds to doubling sensitivity

Azimuth The angle of a tape-head to the tape

Bandwidth The range of frequencies needed for transmission of a signal. The telephone systems requires the range 200 to 2000 Hz, high-quality sound requires about 40 Hz to 18,000 kHz, and a television image requires a band width from DC (zero frequency) to 5.5 MHz

Biasing A system of imposing a steady effect on a variable one so as to use a more linear portion of the effect. On a graph, this corresponds to using a straight-line portion instead of a curved part

Blanket The intermediate image carrier of an offset-litho printer

Bubble-jet A computer-printer technology which uses a matrix of fine hypodermic needles to blow out a fine jet of ink when heated internally

CAD Computer aided design, a method of creating technical drawings with the computer using suitable software. Drawings can be made to scale and can be edited and altered as required

Cardioid The (heart) shape of the response diagram of a microphone which is fairly directional

CCD Charge-coupled device, a solid form of light-sensitive device for use in miniature TV cameras. Both scanning and light-sensitivity are incorporated in the one device

Centronics A form of standard connection for computer printers, using a cable which can be up to 2 m in length

Chrominance signal The colour part of a video signal

Chrome-oxide Chromium oxide, which has superior recording qualities as compared to ferric oxide, but which demands very different settings within a recorder

Colour-cast   The predominant colour that appears on an image due to the colour of the illuminating light, blue in sunlight, yellow in artificial (tungsten) lighting. The eye automatically adjusts for colour casting in real-life scenes, but not in small images

Corona   An electrical discharge through air which makes the air temporarily conductive. It also converts some oxygen into ozone, a gas which causes many materials to deteriorate rapidly

Crystal microphone   A type used on very low-cost tape recorders at one time, obsolete now. Quality is very poor, and the crystal often becomes useless if stored in a warm damp place

Current surge   Resistance of a cold filament (of a lamp-bulb) is very much lower than hot resistance of the working lamp. This leads to a large surge of current, several times the normal working current, when the lamp is first switched on. This current surge is the usual cause of bulb failure

Database   A form of computer program intended to keep information in ordered form, like a filing system

dbx   Type of noise-reduction system for tape which can achieve spectacular reductions in tape-hiss, making tapes almost a quiet as CD recordings

Degauss or deflux   To demagnetize

DX coding   A system of coding a film cassette with notches so a camera can be set for film speed automatically as film is inserted

DIN   A system of measuring film speed in which each increase of three in the number corresponds to a doubling of speed. DIN is also used to denote standard forms of plugs and sockets for electronics use – the initials are of the German national standards organization

Dolby system   A method reducing noise inherent in sound reproduction media. It operates by boosting the volume of treble signals during recording, and decreasing their volume by a corresponding amount on replay. Reduction during replay also reduces noise added in the recording process, greatly cutting the characteristic hiss of a soundtrack using either optical or magnetic tracks

Dynamic microphone   A microphone using an electromagnet and diaphragm. These can be mass-produced very cheaply, with low performance, or made to exacting standards with a corresponding quality and price

Electret   A material which is permanently electrically charged, just as a magnet is permanently magnetised. Electrets are used to make microphones which can achieve surprisingly high quality at low cost

Electromagnet   A coil of wire, usually wound around a core of magnetic material. The material is magnetized when a current flows through the wire. Conversely, a voltage will be induced in the wire when a magnet is moved near the core material, or if the strength of magnetism near the core is altered

Emulation   Of a computer printer, the use of control codes of another make of printer. Among dot matrix printers, the Epson series and the IBM *Pro-printer* are most widely emulated, among laser printers, the Hewlett Packard *Laserjet* is a standard

Emulsion   The layer of gelatine and silver-halide of a film. This is hygroscopic (water absorbing) and will retain fingerprints. On processed film, emulsion is hard, and portions scraped off against the gate of a projector can be quite difficult to remove

Feedback   A return, intended or otherwise, of output signal back to an input. This can result in an improvement in the system if the feedback is negative (opposing the normal input) or in violent oscillation if the feedback is positive (aiding the normal input)

Ferric-oxide   The iron oxide used as the coating for most tapes in the low to medium quality class

Ferrite   A magnetic material physically in a glass form. Used to construct tape-heads, which are very resistant to wear

Fish-eye lens   A lens of very wide angle and short focal length which can show an almost-panoramic view, but with grotesque distortion of perspective

Flutter   Fast variations in pitch of a steady note caused by rapid fluctuations in speed of a tape or disc

Flying-spot scanner   A device for transferring film images to video. Instead of using a conventional intermittent movement, each frame of film is scanned with light, and transmitted light is picked up by a photocell to yield a video signal. Movement of film forms part of the scanning process and can be continuous

Focal length   Distance between a lens and the point at which parallel rays of light are focused

Fogging   Unintended exposure of a film causing blackening

Framing   Appearance of a projected picture with reference to the projector aperture. If portions of another frame are seen, the picture is incorrectly framed, and the film must be moved with respect to the light-aperture

Gate   The region of a movie projector in which film is held and moved intermittently so each frame is projected in turn

Ghost images   Portions of an image appearing under the main image on a screen when a movie is projected. Due to movement of film before the shutter has cut off light

Head-gap Tiny slit in magnetic material of a tapehead across which magnetism affects the tape

Heat-sinks Massive metal fins used to dissipate transistor heat into air

Hum The sound whose frequency is the 50 Hz of the AC mains, or the 100 Hz which is generated in amplifiers when the mains voltage is converted to DC

Hypo Photographer's name for sodium thiosulphate used for removing unwanted silver halide from a processed film or print, thus fixing the image

Infra-red A portion of the spectrum of electromagnetic rays of longer wavelength than visible light. Infra-red beams can be produced by semiconductor lasers and used for remote control and auto-focus actions. Infra-red rays carry most of the heat energy of a beam, so that an infra-red filter glass will greatly reduce the heating of a slide while making little change to the amount of visible light reaching the slide

Interlacing A scanning system in which odd-numbered lines of a scanned image are all scanned before even-numbered lines. This reduces the amount of bandwidth needed while maintaining a rapid frame rate

Laser Acronym of light amplification by stimulated emission of radiation, a process which results in light beams which are conherent (in phase) over long distances, as distinct from the mixture of waves from other sources. The practical result is light which maintains a parallel beam over huge distances and can be focused to a very small spot size. The light is also of one frequency, monochromatic

LCD Liquid crystal display, a form of visual display which produces no light (unlike a cathode-ray tube) but can be used with external illumination. Its main merit is that the display requires very low power to operate

Leader The first (blank) section of a tape or film, intended for threading

Lens formula For a lens of focal length f, an object at distance u will focus to an image at distance v where:

$$1/f = 1/u + 1/v$$

Lithium battery A type of battery in which each cell gives about 3V and which is used when very long life at low discharge rates is required. The battery must never be opened or any attempt made to recharge it, because lithium is a very toxic material which will burn uncontrollably in contact with water, giving choking fumes and leaving alkaline deposits

Luminance signal   The black-and-white part of a video signal

Maltese cross mechanism   The old-established method of moving a film intermittently in a movie camera or projector. Also known as a Geneva movement

Mask-line splice   A splice carried out on the space between frames on a film

Metal tape   A recording tape which uses a metal coating rather then metal oxide for the highest possible audio quality. Metal tape is also used for high-quality video recording, notably for *8 mm* cassettes

NCR paper   No-carbon required (*NCR* is a trademark), a form of copying paper which dispenses with the need for carbon by using chemical coatings on sheets which react with each other when struck

Newton's rings   An interference pattern caused by clamping transparent materials closely together. Light from each of the touching surfaces is reflected, and in places light rays are out of phase, cancelling each other, so that a dark spot is seen. The ring is the 'contour-map' of a complete set of such dark spots

NiCad   The nickel-cadmium type of rechargeable cell. These cells must be recharged in a charger specifically designed for them (constant current recharging) and *never* in the constant-voltage type of recharger used for lead-acid cells

OPC   Organic photo conductor, material used to make copying drums of some electrostatic copiers and laser printers in preference to the rather toxic selenium

Pan   The action of swinging a camera round so as to take a panoramic view or to follow a moving object

Pascal   Unit of pressure equal to one newton of force per square metre of area

Pixel   A picture element, usually taken as being square or rectangular

Plotter   A form of drawing machine which uses a set of fibre or ball pens under computer control

Power of lamp   Measured in watts (W) the product of voltage (V) in volts and current (I) in amperes. A 24V 10A bulb dissipates power of 240W. Conversely, since $I = W/V$, a 250W 40V bulb takes 6.25 amps of current.

Print-through   Magnetization of a tape by an adjacent piece of tape, more common on ferric oxide tapes and occurring mainly when tapes have been stored for some time at high temperature

Pre-echo   Faint occurrence of a sound (usually a crescendo) on a recording before the actual recorded portion, due to print-through

Quarter-wave coating   A coating of a material such as magnesium fluoride on a lens which greatly reduces reflections. Thickness of coating is equal to one quarter the wavelength of light (taking the average of the visible spectrum)

Quartz-halogen lamp   The conventional bulb uses a tungsten filament in an inert gas atmosphere, but the life of the bulb is limited by evaporation of tungsten on to the glass envelope. A trace of a halogen (usually iodine) greatly reduces this effect, allowing the tungsten to be run at a higher temperature. The quartz envelope has a much higher melting-point than glass so can be made much smaller

Radio microphone   A combination of microphone and miniature radio transmitter which allows the wearer to move freely without the need for a microphone-operator to follow with a microphone on a boom. Radio microphone systems are permitted on certain frequencies, subject to strict regulations

Resolution   Ability to display detail. Resolution is usually measured in terms of either the number of distinguishable dots per linear inch (for printers) or as the number of distinguishable lines packed either vertically or horizontally on a TV screen

Ribbon microphone   The highest quality of microphone, often costing several hundreds of pounds and requiring amplifiers of commensurate quality

Rotary-head   A system of recording TV or digital sound signals on tape without requiring the tape to be pulled through the recorder at high speed. Signals to the rotating heads are transmitted through rotating transformers so that no sliding contacts need to be made

Salvage cassette   A cassette which contains no tape, used to receive tape and reels from a damaged cassette

Scanning   The breaking down of an image into sampled spots

SLR   Single-lens reflex, the form of camera in which the image in the viewfinder is the image through the lens

Splicing   Joining of tape or film by means of cement or self-adhesive tape patches. For film, the cut is along the frame line separating two frames, making the splice invisible when played back. Tape splicing can be made inaudible, but only if done with considerable care

Stereo sound   A system of using two separate channels of independently recorded sound to form the illusion of space when the channels are replayed through separate amplifiers and loudspeakers

Striped film   Film coated with magnetic recording tape along the non-sprocketed edge

Thermal fuse   A fuse that responds to ambient temperature rather than to electric current, used to protect equipment against excessive temperatures

Threading   Winding of film or tape through a movie projector, tape or video recorder; also called lacing. Most movie projectors now use automatic threading systems, and all cassette audio and video recorders require no threading as the tape is pulled out into the correct path on loading

Toner   Powder ink for electrostatic copiers and laser printers. Toner has a low melting-point, is magnetic and also electrically non-conducting

TTL   Through-the-lens light metering system for cameras. The initials are also used for a family of digital integrated circuits used in computing and control applications

U-matic   First form of video recording system reliable enough in unskilled hands, and at low enough cost, to be considered for educational use

Variable-area soundtrack   A thin strip of film, opposite the sprocketed edge, which has a variable-width pattern. This pattern represents the shape and amplitude of soundwaves for the film and can be converted into an electrical signal by a photocell. An older method used variable density, using a constant width but varying blackness of film

VHS   Video home system, one of the two original main domestic videocassette systems, designed by JVC

VHS-C   A compact 30-minute miniature cassette system for video cameras. The tapes can be played in a *VHS* recorder if they are enclosed in a suitable casing

VHS-S   A high-resolution form of video cassette system which maintains some compatibility with the original *VHS* system

VU meter   A device for measuring volume units for tape recording – variation of volume units must be kept within a prescribed range to achieve a useful recording

Video signal   An electrical signal that is derived from a TV camera. Shape of the image forms the luminance (black-and-white) signal, and its colour variations form the chrominance (colour) signal

WD40   A proprietary silicone oil which is used for lubricating plastics, electrical potentiometers and a host of other applications

Wow   A low-frequency variation in the pitch of a steady note caused by comparatively slow variations in replay speed of a tape or disc

Xerography   A copying technique making use of the effect of light to discharge materials such as selenium after they have been electrically charged

Zoom   Action of altering the focal length of a lens so that a subject in a picture seems to be magnified or demagnified

Names and addresses

Kodak Ltd provides an emergency health and safety information service which is manned on a 24-hour basis. There is also an advisory service relating to the use of all Kodak products, which should be consulted on all points relating to disposal of Kodak processing chemicals. The address for the advisory service is:

Kodak Ltd,
Product Safety Data Services (A10b),
Kodak House,
PO Box 66,
Station Road,
Hemel Hempstead,
Herts. HP1 1JU
Telephone (0442) 61122 Extension 44538

The 24-hour emergency information service is on the number:
(081) 427 4380
and during the hours of 8.30 to 5.00 Monday to Friday, ask for the Company Health Safety and Environment Department. Outside these hours, calls are diverted by the duty telephonist to the Medical Department or to other available staff.

Other names, addresses and services

APT Radar Systems Ltd (Cybervox Division)
Unit B,
Sprint Industrial Estate,
Chertsey Road, Byfleet,
Surrey KT14 7LA
Telephone: (09323) 51711
Complete language laboratories.

De Vere Ltd
Vulcan Way,
New Addington,
Croydon CR0 9UG
Telephone: (0689) 42222
Photographic enlargers and processing equipment.

Elf Holdings Ltd,
836 Yeovil Road,
Trading Estate,
Slough, Berks.
SL1 4JG
Telephone: (0753) 821886
16 mm and other projectors.

Eric Fishwick Ltd,
Haydock,
St. Helens,
Merseyside.
WA11 0XE
Telephone: (0744) 611611
Photographic equipment, including OHP and slide projectors.

Folex Ltd,
18–19 Monkspath Business Park,
Shirley, Solihull,
West Midlands.
B90 4NY
Telephone: (021) 733 3833
Overhead projection equipment.

Hewlett Packard Ltd,
King Street Lane,
Winnersh,
Wokingham,
Berks.
RG11 5AR
Telephone: (0734) 784774
Microfiche readers.

Kodak Ltd,
Kodak House,
Station Road,
Hemel Hempstead,
Herts.
HP1 1JU
Carousel projectors and photographic equipment.

Kodak maintains a set of special telephone lines for advice on AV
equipment available on a 'local call' basis. These lines are:

Belfast (0232) 328549
Birmingham (021) 643 0733
Bristol (0272) 298121[nl Hemel Hempstead (0442) 61122 or 61241
Glasgow (041) 248 4071
Leeds (0532) 442233
London (081) 427 4380
Loughborough (0509) 218326
Manchester (061) 008 7055
Newcastle-on-Tyne (091) 232741

Kodak Service Centre,
(T.J. Kenyon Ltd),
Bessemer Drive,
Stevenage,
Herts.
Telephone: (0438) 720888

Kodak Parts Services,
Telephone: (0442) 61241
or dial one of the local numbers above and ask for extension 65153

Konica Business Machines Ltd,
6 Miles Gray Road,
Basildon,
Essex.
SS14 3AR
Telephone: (0268) 534444
Photocopiers and other office equipment.

Linguaphone Institute Ltd,
St. Giles House,
50 Poland Street,
London
W1V 4AX
Telephone: (071) 439 4222
Language courses and Minilab recorders.

Matmos Ltd Unit 11,
The Enterprise Park,
Lewes Road,
Lindfield,
W. Sussex.
RH16 2LX
Telephone: (04447) 2091 or 3830
Low-cost computers, computer parts and peripherals.

Photax Ltd,
AV division,
The Gate Studios,
Station Road,
Borehamwood,
Herts.
WD5 1DQ
Telephone: (081) 905 1177
Huge range of photographic and AV equipment.

Rotaprint Industries Ltd,
Rotaprint House,
Honeypot Lane,
London
NW9 9RE
Telephone: (081) 204 3355
Offset litho and other reprographic machines.

RS Components Ltd,
PO Box 99,
Corby,
Northants.
NN17 9RS
Telephone: (0536) 201234
Electronic components, including computer components.

Sony (UK) Ltd,
Pipers Way,
Thatcham,
Newbury,
Berks.
RG13 4LZ
Telephone: (0635) 69500
Video equipment, particularly *8 mm* camcorders and recorders.

Star Micronics (UK) Ltd,
Star House,
Peregrine Business Park,
Gomm Road,
High Wycombe,
Bucks.
HP13 7DL
Telephone: (0494) 471111
Computer printers, including dot-matrix and laser types.

Index